How Science Runs

Eric J. Mittemeijer

How Science Runs

Impressions from a Scientific Career

 Springer

Eric J. Mittemeijer
Max Planck Institute for Intelligent Systems
Institute for Materials Science
University of Stuttgart
Stuttgart, Germany

ISBN 978-3-030-90097-7   ISBN 978-3-030-90095-3   (eBook)
https://doi.org/10.1007/978-3-030-90095-3

Cover Illustration: Image made by Eric J. Mittemeijer from an aluminium sheet owned by him.
Credit: Eric J. Mittemeijer.

This Springer imprint is published by the registered company Springer Nature Switzerland AG
The registered company address is: Gewerbestrasse 11, 6330 Cham, Switzerland

It is not enough to love and admire Nature.
We must understand it, down to and within its core;
that is what Science is all about.

About the Author

Eric J. Mittemeijer was born in 1950 in Haarlem, The Netherlands. He studied "chemical technology", with specialization physical chemistry, at the Delft University of Technology and acquired his "ingenieur (= Ir.)" degree (comparable to a M.Sc. degree) in 1972 and his Ph.D. degree in 1978. From 1985 till 1998 he was full Professor of Solid State Chemistry at the Delft University of Technology. From 1998 till 2017 he was Director at the Max Planck Institute for Metals Research (later renamed in Max Planck Institute for Intelligent Systems) in Stuttgart in conjunction with a full Professorship in Materials Science at the University of Stuttgart. He was Dean of the Study Course Materials Science of the University of Stuttgart and Speaker of the International Max Planck Research School on Advanced Materials. He has (co-)authored about 700 scientific papers in international scientific journals, published a number of books and has received a number of honours for his scientific work. He can be contacted at: e.j.mittemeijer@is.mpg.de.

Preface

Science is not an abstract activity performed by flawless gods. I wrote that in 2010 at the end of the Preface to my book *Fundamentals of Materials Science* (Springer, 2010, 2021). Yet, upon reading a scientific paper or book, never a human emotion, for example of joy, satisfaction or disappointment, will be met. Also the impression is made as if published science is perfect and that scientists are people of impeccable character and morality.

The path of science cannot be imaged as the straight line providing the shortest connection between a question or an assignment and the answer or goal. Only in retrospect such presentations of the development of scientific knowledge can be given. The character and the talents of the scientists and chance determine the path followed. Detours and lateral moves occur. However these deviations from the straight line are not recognized as such while on the winding path to understanding of nature. Sudden flashes of insight and surprising discoveries cannot be predicted and therefore, while underway, it is impossible to be sure about what the best strategy is. Hence, very human character traits decide about what happens in science.

Whereas in many cases scientists act with best intentions to promote progress of science, this is not always the case, because the search for glory and wealth interferes. This can lead to also blunt deceit and fraud. In the last couple of decades such negative eruptions, although of all times, have become very evident and cannot be dismissed as that rare that they can be ignored.

At the end of a lifelong scientific career the true nature of science and scientists, "how science runs", may have become clear. This booklet is about my experiences as a scientist during about half a century. Thereby this booklet is *not* a scientific treatise: it is shown how modern science runs on the basis of also personal experiences with, next to the positive episodes, an open eye for the competition and the nowadays manifest negative developments, as also I myself have witnessed.

To show the personal, human side of scientific research is to demystify the actions of the scientist and to expose the "human drama" on the stage of science. This booklet offers a considered reflection on what matters in science in a hopefully entertaining manner.

I decided to write this booklet after ample consideration. I am a materials scientist, i.e. my scientific work comprises aspects from physics, chemistry and engineering. An aimed for side effect of writing this booklet then is to expose at the same time the soul of materials science, recognizing the relative unfamiliarity with this field of science.

My scientific career can be divided into two parts. The first half was spent as, eventually, a Professor at the Delft University of Technology in The Netherlands; the second part as a Director at the Max Planck Institute for Metals Research in Stuttgart, in Germany, in conjunction with a full Professorship at the University of Stuttgart. These occupations, at different laboratories in different countries, in such a multidisciplinary field of science as materials science, over such long time, may have given me the authority to present my considerations and to recount elucidating experiences in my career such that they are illuminating, interesting and useful to becoming and also older scientists as well as the layman. It was not my intention to write a straightforward autobiography, which, I feel, wouldn't be appropriate.

As can be seen from the titles of the chapters, a number of chapters deal with general, positive and negative aspects of present-day science (especially Chaps. 4, 6, 7, 9, 11 and 12). The other chapters derive partly from personal experiences and run more or less parallel to the course of my career, presenting as well reflections on the history of institutes and relevant scientific developments.

The booklet has been written such that the layman can follow it. It should be of interest for everybody interested in the way scientific research runs at present. Of course, a number of the stories recounted, regarding some specific scientific problems and activities, will have most relevance for materials scientists, physicists and chemists. But these narratives have been presented such that they form no barrier for the interested reader who is a layman in these fields. In a few cases I have added scientific explanation and detail in footnotes, which can be skipped but may be of interest to scientists.

Heidelberg, Germany   Eric J. Mittemeijer
August 2021

Acknowledgements and Justification

I have profited from many of those who interfered with my life. I am immensely grateful to them all. In the foremost place my parents: on former occasions I have overlooked how tributary I am, also as a scientist, to what they invested in me. It is fitting that I express this gratitude now, at the conclusion of my final publication, and it is yet regrettable that I can bring me to such a public acknowledgement only after their death.

This booklet is for a significant part based on personal experiences and thus contains autobiographical elements predominantly connected with my career in science. I am aware that a personal account at the same time uncovers aspects of lifes of other people. This is inescapable, and yet great prudence from my side has been executed to avoid unnecessary or undeserved damaging. Yet, I have not changed names, as I have considered that as unjust, but in a very few cases, usually but not always concerning persons still living today, I have avoided giving names at all; in the latter cases insiders will know who is meant, which is unpreventable.

I am indebted to Mrs. Doris Meissle (Construction Department (= Bauabteilung) of the Max Planck Society, Munich, Germany) for her mediation in the provision of Figs. 13.3 and 13.4, to Mr. Jules Dudok (Science Centre, Delft University of Technology, The Netherlands) for assistance in retrieving and making available Fig. 5.2a and b, and to Mr. Frans van Rumpt (history teacher at the Schoter Scholengemeenschap, Haarlem, The Netherlands) for making, on my behalf, the photograph shown in Fig. 3.1. My colleague Prof. Paolo Scardi (University of Trento, Italy) kindly provided me with a copy of Fig. 10.2. I am grateful to my former Ph.D. student and now Prof. Marcel Somers (Technical University of Denmark, Lyngby) for refining my memory regarding some details of the story recounted in Sect. 9.3. My former co-worker Dr. Ralf Schacherl (University of Stuttgart) prepared final versions of a number of the figures. My brother Kees searched and found in the possessions of our deceased mother the photograph shown in Fig. 1.1, that I had last seen about 60 years ago, but which image was highly visible on my mind. I also thank Dr. Zachary Evenson (Editor Research Publishing—Books, Physics & Materials Science, Springer-Nature) for valuable advice and support in the preparation of this booklet.

Upon reading the Epilogue to this booklet, it will be clear how important my wife Marion has been for me in the second part of my life. She did not agree that I consumed time on drafting this booklet, as this robbed a lot of time that we could have spent together. The more my gratefulness to her increased even more, as she let it happen. Moreover, she provided constructive criticism of text parts of the manuscript.

The Corona pandemic has constrained the travelling activities of the two of us, making staying in our holiday apartment in Italy impossible, and this enabled completion of this booklet as well…

Heidelberg, Germany   Eric J. Mittemeijer
August 2021

Contents

1

The Parents

Abstract A brief description of my immediate ancestry.

During my parents first contact any visual impression was impossible. George Mittemeijer worked as a clerk in the post office in Amsterdam; Annie van Campen was employed at the telegraph office in the same city. For some reason a professional call to the telegraph office had to be made by George and by chance the person he contacted was Annie. The voice of George was warm and appealing and that sufficed to arrange a first date with Annie (see Fig. 1.1).

Annie and George were affianced for quite a number of years, six, before they at last married in January 1949. Of course, the Second World War and the meagre years thereafter in many ways made life difficult, but this could not be the major reason for a long engagement. One obstacle to get married certainly was that Annie's father withheld his consent to this marriage; for marrying at an age under 30 approval by the parents was necessary by law in The Netherlands until 1970…. So George and Annie went to court in 1948 (they were 30 and 23 years old, respectively) and formally got permission to marry from the cantonal judge. This experience for Annie came on top of the earlier refusal of her father, against both the strong advice of her teachers (Annie had highest possible marks for mathematics) and her own vigorous desire, to let enroll her on a higher secondary school, which would have allowed her to enter university thereafter. Annie never met or contacted her father again.

E. J. Mittemeijer, *How Science Runs*,
https://doi.org/10.1007/978-3-030-90095-3_1

Fig. 1.1 Annie and George at the day of their marriage at the 15th of January 1949

George father's family was of German origin, but at some stage had managed to write the family name in a more Dutch manner: "Mittemeijer", instead of "Mittermaier" and variants thereof.[1]

George had completed a short 3 years version of the HBS (a secondary school; see Chap. 3), which was actually considered as basis for a following 2 years commercial school ("Handelsschool") education, which, however, he did not follow. The unfavourable economic situation (the Great Depression,

---

[1] Replacing, in the current fashion of writing the family name "Mittemeijer", the first part "Mitte" by "Mittel" or "Mitter" and/or the second part "meijer" by "meier" or maier", variants of the family name occur that are currently still known in Germany. That suggests that, in the past, an ancestor of the "Mittemeijers", with an in the above sense modified, closely related family name, moved from Germany to The Netherlands.

The progenitor of all the "Mittemeijers" in The Netherlands is "Johann Georg Mittemaier", who was born in 1762 in Swaben, at the time a small village, now incorporated as a part of Waldenburg in Saxony ("Landkreis Zwickau"). At some time between 1762 and 1790 he must have moved to Amsterdam in The Netherlands, where he married "Helena Sophia Roer" in 1790 and died in 1835. Apparently "Johann Georg Mittemaier" must have adapted the way of writing his name in the course of time after his move to The Netherlands. He may have even been illiterate (to some extent; certainly not unusual in the eighteenth century). His first and family names are written in various ways in documents of the time. In the notice of marriage dated 23-4-1790 it was written: "Johan Georg Mittemaijer". In the later birth certificates of his children his first names had sometimes been indicated as "Jan George" but mostly as "Johan George", whereas the family name now definitively (?) had become "Mittemeijer". I descend in the seventh generation in a straight line from him. The male ancestors had professions as blacksmith, shoemaker, mechanic and, finally, office clerk (my grandfather).

lasting from 1929 till 1939) made him start his professional life as an unpaid apprentice ("volontair"). Soon thereafter he acquired a paid position at the post office where the above described first contact with Annie was established. The job at the post office allowed George a lot of contact across the counter with clients. Shortly after his marriage he moved to an administrative position at the central bank of The Netherlands ("De Nederlandse Bank") which job was, important for the young family, better, yet still modestly paid, but without any personal contact with customers. This man-to-man, human contact was dearly missed. Therefore his early retirement, an indirect consequence of a very long lasting, late diagnosed tuberculosis ailment, which cost him a kidney and a lot of eyesight, was considered something as a relief.

A well-balanced personality characterized George and in this way he was an ideal partner for Annie. He was also humorous and he was liked by many within and outside the family. I remember him singing funny and also scandalous songs, which I can still reproduce to the hilarious delight of my wife.

Upon their marriage George and Annie moved to Haarlem where they rented a basement two room apartment where their two sons were born (Eric in 1950 and Kees in 1952).

A few years before the 50th anniversary of their marriage my father suffered a cerebral hemorrhage in 1997, while I was on the verge of moving to Germany. The same day I hurried to and visited him in the hospital. He seemed to recognize me but could not produce anything else than non-understandable sounds: the brain was severely damaged. He died in the following night at the age of 79.

My mother survived my father for many years. In the concluding years of her life pronounced health problems prevailed, which made life uncomfortable and very painful. Her life ended in 2019 at the age of 93. Until the very end she had full command of her intellectual power; not the slightest sign of dementia she had shown.

2

Growing Up

Abstract Life in the 1950's. Early experiences; growing up under intellectual stimulation. Good and less good primary schools. Marvelled by the sound and music a piano produces.

The Tetterodestraat in the Kleverpark quarter in Haarlem in the fifties was a quiet, cross street with lines of older, relatively large, usually two-storey houses at both sides; no gardens in front of the houses and no trees[1] to enliven and cheer up the rather uninspiring atmosphere confronting the passer-by. Mostly lower middle-class people lived here. Often more families lived in one house, each occupying one floor.

After the Second World War an enormous housing-shortage ("woningnood") occurred in The Netherlands.[2] So my parents were happy to rent the ground floor of the house at Tetterodestraat 16. As compared to present conditions, living there for a family of four persons took place under severe constraints. The apartment consisted of two rooms: the living room and the sleeping room for the parents. The children slept in a small wintergarden adjacent to the sleeping room. This wintergarden had both a glass ceiling and a glass front to the garden behind the house. Therefore sleeping in winter there was occasionally an extremely cold experience. The only place for the

---

[1] According to my memory and in contrast with the current situation.

[2] This housing shortage, more precisely lack of houses/apartments *to rent*, prevailed far into the seventies. The currently experienced housing shortage cannot be compared with the one after the Second World War, as the present one is merely the consequence of a growing number of single households and the need of much more square meters per person.

family to wash themselves was a single wash-hand basin in the sleeping room, delivering of course only cold water. In summer a tub was placed in the garden with warmed water where the kids were washed…

The floor of the living room was covered with coco matting, the one of the sleeping room with balatum, both being cheap floor coverings. The heating was realized in only the living room by a rather small stove fed with coal. A small radio stood on a side board. In the fifties television did not commercially exist and a refrigerator and a washing machine were much too expensive.

From the above description a current observer looking at the family Mittemeijer in the fifties might wrongly say that the family was poor. Admittedly, my father had an only modest income; in about 1960 it must have been of the order 6000 guilders (say, 3000 Euros) per year, but this was an acceptable income for those days. The point is: at the time Holland was not a welfare state.

Those of us who were born around 1950, as the author, have experienced during their lifetime the steady, enormous increase of prosperity and technical facilities making daily life so comfortable. This evolution, incorporating also some revolutionary jumps forward (as birth of the world wide web), has made a deep and lasting impression on me; I cannot forget where we came from.

Upon marriage my mother had to give up her job[3] and thus stayed at home as a housewife. Only decades later it became normal for a married woman to have a job. My mother didn't like to be without a job: a two room apartment and two kids who stayed during the day at Kindergarten/school do not imply a day filling occupation. She longed to be back at the telegraphic office and indeed later returned (accepting a full job after her sons attended secondary school) and made a career eventually becoming a head of department. She certainly was an ambitious person. Modest trips became possible: a VW beetle was bought and the family for the first time made holidays in the mid-sixties in the sense that a journey abroad, to Austria, was made.

Even as a small child I was given certain responsibility. Each day I had to go by feet to the Kindergarten. This implied I had to cross also a major street, with a significant amount of traffic. The first times my mother went with me and trained me how to cross that street safely. The next time she went with me to that street and let me cross the street by my own, overseeing what I did. Thereafter she believed correctly I could do it alone, unwatched by her.

---

[3] Women in government jobs and younger than 45 (as Annie) were obliged by law of 1924 to give up their job at the moment they married (many years earlier Annie's mother had suffered the same fate). It would last until 1958 that the government abolished this prohibition, only after a long struggle of especially female activists with male catholic and protestant politicians who dominated in this confessional and at the time in a number of respects arrantly conservative country.

In this way she had taught me, at a very young age, something very important: to act independently and be responsible for myself, i.e. also carrying the consequences for one's own actions. And of course it felt good to be trusted. Annie and George were not at all "helicopter" parents: in no way they would have ever thought to bring me personally with a car to school, if they would have had such a vehicle.

I was a rather curious child, trying and wishing to understand my "world". I saw my parents reading the newspaper and wished to know what they actually were doing. After they had (more or less) explained it to me, I insisted on wanting to learn reading myself. Thus, at the age of 4 or 5, my mother taught me the alphabet and I could read to significant extent before I went to primary school. She also taught me counting and before I entered school I could perform simple additions, subtractions, even multiplications and divisions.

The atmosphere at home was bent on very positive response to expressed willingness/wish to learn, which one can call discovering the world. My parents, in particular my mother, did everything to stimulate questioning behind what one saw or experienced. This is nothing else than transferring what should be a fundamental attitude of a scientist; the beginning of what is called science, of which I was obviously totally unaware at the time. Years later, when I attended secondary school, the situation was reversed: my mother, in view of her education on a lower level and urged by her interest in and talent for mathematics (see Chap. 1), didn't know what goniometry involved. So I explained to her what a sine, tangent etc. are and what type of calculations can be performed based on these mathematical functions. She understood.

For some years the children with the parents in the very early morning went to (the public) Stoop's Bath in Overveen, which was at walking distance. My mother had decided that we should learn to swim. She herself was a good swimmer. Both children thus got their swimming certificates.

It was in the swimming pool that my father described and demonstrated Archimedes' principle of buoyancy to me. He was not well trained at all in physics and, as I realized much later, certainly he could not have presented a derivation of this principle, but he had absorbed something of significance from what was taught to him in his short school career.

My parents, both Annie and George, each within their means, did not wait for and rely on only (primary) school to trigger intellectual curiosity in their children and made learning for them a natural, essential and above all pleasant, playful part of life.

At the age of six I went to the Tetterode school, a primary school in an old building at the corner of the Tetterode straat, the street where we lived, and the Santpoorter straat. This was a very good school: it was called an "opleidingsschool", i.e. a school preparing for secondary school on the level of e.g. a gymnasium. From the first day I enjoyed being there. The education was certainly somewhat old fashioned there: so-called school boards were used and I also had to learn to write with a crown pen to be dipped in ink. But, these are only superficial characteristics, which tell nothing about the quality of the teaching (also see Chap. 3).

I attended this school only two years. One memory stands out. In the second class, at the age of seven, each pupil had to tell in front of the class about what it had read or experienced, for example during holidays. As it turned out, I just had finished reading a children's booklet called something like "The mysterious neighbour". The story, as I remember it, likely in more or less distorted manner, after more than 60 years, is about a couple of children who move with their parents to a new home. Soon they find out that next to them a neighbour lives who behaves strangely. They decide to investigate it and are confronted with a series of peculiar occurrences, such as that audible life happens in the house of the neighbour also in the evening and night, although the house appears and remains completely dark, no light seems to shine inside. There is an atmosphere of suspense gradually building up and tension and fear developing at the moment they go into the garden of the neighbour to have a closer look. The narrative then experiences its "chute": suddenly the neighbour appears in the garden and moves about in a rather singular way, wearing a pair of spectacles with dark glasses. Then, abruptly, the children realize that their neighbour is blind and they understand everything. They have nothing to be afraid of and become friends with their neighbour. Apparently I was able to recount the story such that I could make felt the growing uncertainty of the investigating children and their alarm upon the appearance of the mysterious neighbour in the garden. So that the subsequent recognition of the neighbour's blindness was sensed by my audience as a relief, as I had experienced myself while reading the booklet. This was the first time I was capable to bind the attention of a group of listeners and that without having felt any shyness in standing in front of an audience, although I was, and in fact still am, a rather shy child/person. Although I didn't recognize it as such at the time, it must have been a primal experience for me. In my later life as a scientist and Professor I have given countless lectures at conferences and to students. I did that with great pleasure, felt comfortable as a Speaker/Lecturer and I believe I was capable to draw and keep the interest of my audiences.

My parents with their two sons and the rather often presence of my father's mother, who had serious difficulties to live alone after the death of her husband briefly after Eric's birth, realized that living in only two rooms was not a permanently acceptable situation. In those days home exchange, restricted to rented apartments/houses, was a usual means for finding more appropriate living conditions for the parties swopping their homes. Newspapers published many advertisements of people looking for a party to swop homes. Such home exchange often was not restricted to two parties, but a series of home exchanges could happen incorporating three till, say, five parties (hence in the latter case: A → B, B → C, C → D, D → E and E → A). So, for example a couple, after their children had moved out and lived elsewhere, could find a smaller apartment/house by exchange with a couple with young children looking for an apartment/house with more space. Especially if more parties than two are involved a lot of organizing work is required. In this way my parents were able to move, by home exchange with three parties, to an entire house with six rooms in the Jan van Walréstraat 8, at the corner with the Vosmaerstraat, in the eastern part of Haarlem, in the middle of a working-class neighbourhood. At the Jan van Walréstraat–Vosmaerstraat street crossing a greengrocer shop and a dairy shop occupied opposite corners. A little further in the street there was a butchery. All these shops have now disappeared, likely owing to competition with the emergence of ever more supermarkets in the sixties and beyond.

The move to the house in the Jan van Walréstraat of course was an enormous improvement, as the children got their own rooms. There was, however, at least one serious disadvantage: the neighbourhood, Haarlem-Oost, was of lesser standing then the Kleverpark quarter and it was clear to my parents that they would return to the Kleverpark neighbourhood as soon as that was possible.

A further distinct disadvantage was that I had to visit a local primary school, the van Zeggelen school. This school was not of the same level as the Tetterode school. I was about one half year in advance of my co-pupils upon entering this school. This continuously worried my parents during the years I went to school here.[4] In these days one had to pass an admission examination

---

[4] Another, less serious, disadvantage, as side effect of my change of school, concerns its consequence for my handwriting. At the Tetterode school I had learned hand writing with so-called "block" characters. This was one of those "modernizations" in education lacking a rigorous scientific basis (cf. the comment in Chap. 3). At the van Zeggelen school I had to change to "connected upright writing". This inconsistency in learning to write clearly and neatly by hand damaged incurably my handwriting. Thus, many years later, my students often had to ask my secretary, who naturally was very familiar with my amputated way of hand writing, what I had scribbled in the margin of a manuscript.

at the end of primary school in order to be allowed to enter a HBS (see begin of Chap. 3) or gymnasium. So my mother during the sixth and final year of primary school bought a booklet full with problems of previous admission examinations and trained me. She even approached the teacher and insisted that this type of preparation should also be done by the entire class. Only at a very late stage of the year the teacher handled a few of these problems with the class, strikingly using the same booklet with problems my mother had bought for me. In the end I and only one other pupil of my class managed to pass this entrance examination....

Damage a School Can Cause

There is a lesson to be learned from the above for primary schools in so-called "problematic" or "less favourable" neighbourhoods: teachers at such schools tend to adapt the level of their teaching to the level that the average, or even the less talented pupils can just process successfully. At first sight this seems a very human and social approach, but only at first sight: the unwillingness of teachers to teach at such schools at the level required for entering successfully a gymnasium or similar, as they should do, in classes where indeed and admittedly the vast majority of the pupils may not have the qualities required for attending such education, causes severe damage to our society, because this inexcusable attitude of focusing on the majority of less talented pupils can kill the possibility of entering such higher secondary schools for the few pupils in such classes who have the talent for that but who, apart from possibly remaining unnoticed, become inadequately trained.

The above abuse is worsened by some boundary conditions nowadays prevailing in some countries: (i) where it appears to be forbidden that a pupil in a class in primary school is required to repeat a year in case of insufficient progress in learning, i.e. also such pupils are promoted yet to the next higher class, and (ii) where no admission examinations for secondary schools are required anymore, and instead the wish of parents dominates the choice of secondary school independent of its appropriateness for the child concerned.

A sister-in-law of my father's mother, Fie, lived in Nieuwendam, a former village, now part of Amsterdam, situated at the north bank of the "IJ", a wide water that once was a sea arm of the Zuiderzee. One could reach Nieuwendam by crossing the IJ with a ferry, starting from a quay at the rear of the central railway station of Amsterdam. For a small child the trip with the ferry was already an adventure, as the water of the IJ was not always very smooth and there were a lot of also huge, seaworthy ships, to me seemingly on collision course with the ferry, on their way coming from or going to the North Sea through the "Noordzeekanaal" (north sea canal) that connects the North Sea and the IJ. The house of Fie stood on a dyke. She had an electricity supplies shop. However, the attraction of the visit to me was the piano in the living room. It was the first time I was directly confronted with such an instrument. I marveled at the sound produced upon pressing its white and black keys. From that moment on I wanted to learn to play piano. This repeatedly expressed, sincere wish at last convinced my parents and a few years later a second hand piano was bought and placed in our house in the Jan van Walréstraat. The piano teacher was Cor Rackwitz, who lived in the Vosmaerstreet not far away from us. Piano playing became a great love of me for the rest of my life.

3

Touching Science; School Years

Abstract First contact with science on the Lorentzlyceum; connections of the namesake, Hendrik Antoon Lorentz, with my (scientific) life. Notes on a textbook for physics, the roles of "nature" and "nurture" and the usefulness and uselessness of school collectives and joined first-year classes ("brugklassen"). Death of my piano teacher.

The "hogere burgerschool" (= Higher Civic School; abbreviated: "HBS" or "hbs") was founded in The Netherlands in 1863 and since 1917, as a five year course, it was one of two secondary school forms allowing, after passing a final written and oral examination, access to a study at a university. The other such secondary school was the much longer existing "gymnasium" (= (more or less) grammar school). The progressing industrialization and increasing prosperity in the second half of the nineteenth century increased the need for qualified engineers, manufacturers, bankers, tradesmen, civil servants, and more. The gymnasia did not qualify for meeting these needs.

In a short time the HBS became an enormous success. The education it offered had, strikingly different with the gymnasium at the time, a strong emphasis on mathematics, physics, chemistry and modern languages. It was obliged to be taught in both English and German and French, apart from being taught in Dutch, of course; no lessons were given on Latin and Greek. Its teachers generally were of high quality, usually had accomplished an academic education and often had completed a dissertation (i.e. had performed scientific research). It is noteworthy to mention here, as an example, that van der Waals, a Dutch physics Nobel prize winner (1910),

E. J. Mittemeijer, *How Science Runs*, https://doi.org/10.1007/978-3-030-90095-3_3

had been for considerable time a teacher at precisely this school type. The very most of the Dutch Nobel prize winners originate from an HBS education. Some examples in the fields of physics or chemistry are: in the more remote past: Van 't Hoff (1901), Lorentz (1902), Zeeman (1902), Kamerlingh Onnes (1913), Debye (1936) and Zernike (1953), names known to every physicist and chemist; and from more recent times: Crutzen (1995), Veltman (1999) and Feringa (2016).

It was never an option for me, after primary school, to continue my intellectual education at a gymnasium. The main reason was that I did not see any distinct advantage for me in learning Latin and Greek, which would cost an extra year (the gymnasium had a six year, instead of five year course). It is different of course for a (later, university) study where knowledge of such languages is a prerequisite, but learning Latin and Greek is a useless expenditure of time with a view to application/need in daily life and for studying natural sciences. And of course, it had not escaped my attention that the HBS had an excellent reputation. My parents saw it the same way. So in August 1962 I began my HBS "career" at the Lorentzlyceum[1] in Haarlem, where I would follow the HBS-b direction. The choice of the Lorentzlyceum was also based on the wish and anticipation of my parents to move back to the Kleverpark quarter, which indeed happened a couple of years later.

The Lorentzlyceum was situated in a nice yellow/grey stone building presiding over the Santpoorter plein (= square) in the Kleverpark quarter in Haarlem. The two-story building stretched over an entire, smaller side of the square and had three adjacent gates with rounded upper sides as main, grand entrance at the side of the square (Fig. 3.1). The pupils had to use the less appealing back entrance, of course.

On Lorentz and Points of Contact; the Lorentz Factor

At this place some attention to the namesake of the Lorentzlyceum in Haarlem is in order.

Hendrik Antoon Lorentz (1853–1928) was one of the greatest scientists of his epoch. As a theoretical physicist he was in high esteem of his colleagues and can be considered as one of the leading physicists of his time, where his chairmanship of the first five famous Solvay Conferences (1911–1924) played an important role. His reconciling personality and his command

---

[1] A lyceum incorporated an HBS and a gymnasium. After one first year to be followed by all pupils one had to decide for either the HBS or the gymnasium course. Further the HBS had an "a" branch and a "b" branch. The "b" branch, HBS-b, was the course resembling the original HBS as sketched above, i.e. with the focus on mathematics, physics and chemistry and the modern languages; the "a" branch", HBS-a, was considered to be somewhat "lighter", with a focus on economy and modern languages.

Fig. 3.1 The Lorentzlyceum at the north end of the Santpoorterplein in Haarlem (*Photograph by Frans van Rumpt*)

of foreign languages supported him in acquiring this internationally highly visible status. Albert Einstein often visited him and, nearing the end of his own life, wrote of him: "*For me personally he meant more than all the others I have met on my life's journey*". Work by Lorentz can be considered as paving the way for the special and general theories of relativity. But he has done much more (see the recent biography by *F. Berends and D. van Delft, Lorentz, Prometheus, Amsterdam, 2019* (unfortunately in Dutch) and see what follows).

Apart from Lorentz just having lent his name to my school, there are a few special, also more personal reasons for me to dwell upon Lorentz here.

Lorentz was Professor at the University of Leiden from 1877 (at the age of 24) till 1912. With the passing of the years he felt that teaching, management and his increasing international responsibilities consumed too much time, i.e. withheld him from own research. So he accepted in 1909 the position of "curator" of the "Natuurkundig Laboratorium" (= Physics Cabinet) in the Teylers Museum of the Teylers Foundation in Haarlem. This position provided him in Haarlem with a laboratory and funds for research. After 1912 his ordinary professorship in Leiden was changed into an extraordinary one; he kept giving his famous Monday morning lectures there (see "*Professor Burgers*" in Chap. 10). As the result of these changes he moved to Haarlem and stayed there until his death in 1928. During these years his international status increased more and more. His funeral in Haarlem became an act of national mourning in The Netherlands. Thousands aligned the streets where the funeral procession passed. Ernest (Lord) Rutherford, Marie Curie, Paul

Langevin, Albert Einstein attended the funeral, as well of course his successor in Leiden, Paul Ehrenfest; they honoured the deceased with moving speeches. Haarlem and The Netherlands had not seen before and have not seen thereafter such respect and honour bestowed on a scientist. Lorentz was buried in the Algemene Begraafplaats (= general cemetery) at the Kleverlaan, which is very close to the Lorentzlyceum.

Of course, during my stay at the Lorentzlyceum I was taught about the "Lorentz force" in the physics lessons. However, the much broader importance of Lorentz for modern physics was not exposed to me. I do not remember that a single sign (as for example a plaquette) within the Lorentzlyceum made the slightest reference in some way to this great man. The overwhelming impact Lorentz has had on the scientific world of his time and on his fellow countrymen was not recounted. I learned about that only later in life, at the university.

I have visited Teylers Museum in my younger years various times; the last time, in likely about 1971, when I had organized an excursion for and with fellow physical chemistry students from the Delft University of Technology. The museum is devoted to natural history. It is one of the most interesting and beautiful museums of this kind in the world. Of course the famous, enormous electrostatic generator (in Dutch: "elektriseermachine") built upon instigation by Martinus van Marum at the end of the eighteenth century, that can produce a potential of 300000 V and thereby generate sparks of more than 60 cm, catches the eye. But there is a great lot to see and consider carefully: (reconstructions of famous) scientific instruments (as Foucault's rotating mirror method for measuring the velocity of light), spectacular fossils and minerals, and many drawings by famous artists, as Rembrandt, and much more. One feature of the museum is a beautiful lecture hall with an atmosphere and style of (now) a century and more ago. I remember the names of famous scientists of the (remote) past reproduced on the walls of the hall. Standing there I dreamt to once give a lecture in this lecture hall. It never happened.

Lorentz did not go out of the way of practical problems and felt responsible towards society. Thus he was asked by the Dutch government to be the chair of a committee to calculate the effect on the water levels of the proposed (flood control) dam, "Afsluitdijk" (= enclosure dam), to close the Zuiderzee from the Waddenzee (= Wadden Sea) and thus the Noordzee (= North Sea). In the end Lorentz had to do the hard work almost on his own, especially involving developing the appropriate, approximative theory. It took an enormous amount of his time; by dogged perseverance he worked on it from 1918 till 1926. The numerical calculations were performed together with a dedicated co-worker. The "Afsluitdijk", with a course based on Lorentz's calculations, was completed in 1932,[2] after his death, with resulting effects on the water levels as largely predicted by Lorentz. This story bears a lesson: there is nothing wrong and it is rewarding for a scientist, even a theoretical scientist, to also work on a practical, engineering problem. This is mirrored in a substantial amount of my own work.

The significance of the legacy of Lorentz for science is also reflected by those phenomena and theories in physics that bear his name, notably the *Lorentz force*, the *Lorentz transformation* and the *Lorentz contraction*. In the literature one also recognizes the *Lorentz factor*. However, here ambiguity arises: as a matter of fact there are two Lorentz factors to be encountered in the literature, which have nothing to do with each other. Both factors originate from work primarily done by Lorentz and therefore are associated correctly with him. Search with a search engine in the world wide web ("internet") using "Lorentz factor" as search term provides many links to only one of these Lorentz factors: the one that describes the (relativistic) changes of time, length and mass of an object that is moving with a certain, constant velocity. The other Lorentz factor evidently is much less known. It was obtained by Lorentz in the time he lectured about the recent discovery (in 1912) of X-ray diffraction by crystalline material. Lorentz showed that the diffracted intensity is proportional with the scattering volume and that thereby the geometric, measurement-technique related constraints explain the strong decrease of intensity upon increasing order of reflection/upon increasing diffraction angle. This Lorentz factor plays a distinct role in the quantitative analysis of (X-ray) diffraction experiments (e.g. allowing the determination of crystal structures (i.e. the regular arrangement of the atoms in crystals); see the intermezzo "*Diffraction Analysis as a Tool for Determining the Constitution of Matter; Regularities and Irregularities of the Atomic Arrangement*" in Chap. 5) and therefore is well known in the fields of crystallography and materials science. Research in these fields of science and thus X-ray diffraction as a method of analysis have played a large role in my scientific work and this explains why attention to this in general lesser known, second Lorentz factor is paid here (using "Lorentz factor + diffraction" as search item in the world wide web does yield a list of links related to the second Lorentz factor).

There is a further reason to consider the history of the second Lorentz factor. In spite of its pronounced importance, as indicated above, Lorentz himself apparently did not consider his result that important that he spent

---

[2] Thereafter the Zuiderzee was no longer a sea (= zee), but a lake, and it is called now Ijsselmeer (lake = meer).

time to publish a paper about it. Instead he forwarded this result in a letter to Peter Debye, who published, referring to Lorentz of course, the result in a paper by himself in 1913. In these days pressure on publishing was not as extreme as it is nowadays. This may partly explain that Lorentz did not himself present this second Lorentz factor in a paper. I cannot imagine any scientist of nowadays who would not have gone at significant length to publish a result as the second Lorentz factor by him(or her)self....

The first two years Eric went each day to the Lorentzlyceum by bus, as he had to go from the south-eastern side of Haarlem (Haarlem-Oost), where he lived with his parents and his brother, to the more north-west located part of the city, where the Kleverpark neighbourhood was situated.

The bus passed through the Hannie Schaftstraat named after a young militant, female resistance fighter during the occupation and repression by the Germans in the Second World War. Hannie Schaft, "het meisje met het rode haar" (= the red-haired girl), was executed, in the dunes at the sea, close to and west of Haarlem, at the age of 24 in April 1945, only a few weeks before the war ended. Remarkably, as a coincidence with Eric's school career, she had attended the Tetterodeschool as well as the HBS-b at the Lorentzlyceum.[3] In her case (see the above intermezzo about Lorentz) a plaquette in the hall of the Lorentzlyceum did commemorate her courageous life and deeds.

As there was no time enough to return to home in-between the morning and afternoon lessons at the Lorentzlyceum, Eric went to his grandmother (his mother's mother) and his great-aunt Narda, who at the time lived in Bloemendaal, a village west from Haarlem and on moderate walking distance from school. Although Eric's grandmother and her sister were vegetarians, they insisted on preparing his sandwiches also with slices of sausage and, especially, sardines laid in oil. Eric hadn't tasted sardines before and became very fond of this small fish prepared in this way. Even now, after so many years, Eric cannot eat or even only smell (a sandwich with) sardines without awakening these vivid memories of lunch with his grandmother and great-aunt during these first years of attending the Lorentzlyceum.

---

[3] This is not the only parallel with me. I also had red hair (now much less intense and turning grey). Further, as I found out recently, in the written record about Hannie Schaft's time at the Lorentzlyceum it is mentioned that she was very ambitious, that during her school career she was the best or one of the best of her class and that she was no person with prominent behaviour among her classmates. All these characteristics held for me as well. I have long hesitated with mentioning this here: I do not want to be accused of believing in supernatural phenomena; I do simply find all these coincidences, from school career, a hair colour rather rare in The Netherlands, to character and intellectual characteristics, funny and consider them as the result of sheer chance, but remarkable enough to indicate them here.

My grandmother and her sister took great care and exerted pressure such that I always ate everything that was put on the table during those lunchtimes described above. My mother had the same attitude with respect to the meals at home. It was considered highly immoral to throw away food. I have no doubt that this had to do with Second World War experiences of enduring impact:

The last winter of the war, 1944–1945, was a horrible one for the western part of the Netherlands that is "Holland"; the southern part was already liberated by the allied forces.[4] This last winter of the war was characterized by great famine for the population; people were starving. This winter is called the "hongerwinter" (= hunger winter); every Dutchman is familiar with this expression. My great-aunt Narda used to tell that one day she was on the street and a car with German soldiers passed along. Her hunger and anger were that strong that she, for a short moment fearless, chased the car running while shouting: "Honger, honger!" As a response one of the German soldiers threw half a loaf of bread on the street in her direction. Especially the city dwellers suffered. Long tracks of people walked to the country side hoping to acquire food from the farmers, who generally did not suffer from food shortage. Many farmers helped; some not. My mother took part in these tracks; she rarely spoke about that.

The atmosphere surrounding my grandmother and great-aunts differed very strongly from that at home. The floor in their house was covered with carpets, thick curtains hung before the windows, the rooms were full with couches and large and small tables; precious vases, small sculptures cut from wood and bone and a beautiful wall clock served as decoration. There were always vases filled with flowers, usually roses. Some of the pictures on the walls were attractive to me: even now one engraving (by Voskuyl) and one charcoal drawing (by my great-grandfather), former possessions of my grandmother and her sisters, decorate a wall in my office at home. A certain distinctly civilized atmosphere prevailed in the surroundings of my grandmother and great-aunts. I liked it and felt comfortable there. They possessed an impressive amount of books. Especially my curiosity arose upon noticing that some books, in particular by famous German authors as Goethe and Schiller (in the latter case his complete works), were printed in gothic font.

---

[4] The combination of the present-day provinces North-Holland and South-Holland, i.e. the western part of The Netherlands, is the former county "Holland". The largest cities, Amsterdam (capital of The Netherlands), Rotterdam and The Hague (seat of the government), are situated there. Often "The Netherlands" are indicated with "Holland", also by Dutchmen, which, however, is incorrect and historically in any case misleading.

I had never seen that and couldn't read it, but my grandmother and great-aunts, and also my mother as I later found out, could. This part of my family had indeed read widely, but lacked a basis in natural sciences.

My great-aunts, but not my grandmother, had affinity for what I would call forms of mysticism. They believed in supernatural forms of being and had been, for at least considerable time, votaries of the Rosicrucian movement. As a young boy this all intrigued me highly and I was certainly not insensitive to it. However, I soon realized I was in fact an agnostic and have remained that during my life.

Eventually, after the death of my grandmother, who survived all my great-aunts, I inherited all books and the pictures mentioned, including the literature on the Rosicrucians, as no other member of my family had any interest.

The strong wish of my parents to return to the Kleverpark neighbourhood in Haarlem could eventually be realized in 1964, again by home exchange (see Chap. 2). Thus we moved to the Aelbertsbergstraat 51. Like the Tetterode-straat, the Aelbertsbergstraat is a quiet cross street, but it makes a somewhat better impression, also because most of the houses in the street, as also our house, had a small garden in front of the house, in addition to a large garden at the back of the house. From the house to school it was only a few hundred meters, so from now on I went to school by feet.

After a few years my parents were offered the possibility to buy the house they lived in. The owner until then was a classical rack-rent landlord who had become old and tired of handling the complaints of his renters concerning house deficiencies which had to be repaired. As the house was in rented-out state, the price of the house was considerably lower than if the house could be offered in inhabited state to a new owner. The price was about 11000 Dutch guilders (about 5500 Euros), an indeed relatively small sum for these days (the house was large, in acceptable condition and in a beloved neighbour-hood) and an incredibly small sum as compared to house prices of today. Yet my father much hesitated. He believed, and said once, that "*we are not the kind of people to possess a house*"; he feared the burden of the necessary mortgage and was horrified by the idea to be in debt. However, he could be convinced that he was well in the position to buy the house and it also helped that his employer, the central bank of The Netherlands ("De Neder-landse Bank"), offered very attractive conditions for the mortgage. It was the best decision of his life, as it assured a financially worry-free renter life for him and his wife. Some 15 years later, the mortgage repaid, my parents made their sons owner of the house, i.e. transcribed the house on Eric and Kees, as the way to avoid that their sons would have to pay (considerable) inheritance tax on the house in due course. The notarially fixed condition was that my parents could live in the house until their death. And so it happened.

Looking back to my HBS-b years spent on the Lorentzlyceum, it can be said that these years prepared me well for my later university study; the Lorentzlyceum indeed was a very good school.

I was impressed by my mathematics teacher, (Jaap) van Wamel, who could introduce us to specific topics in mathematics in a very well structured way. He almost did not use a textbook, but his notes on the blackboard and his explanation were excellent and practically sufficed to learn what was required. From these days I learned to work out at home the notes I had made during the lesson, write it up in my "own" words, and doing so actually acquired genuine insight: I had to understand what I wrote. This at first sight somewhat laborious but in the end very efficacious system of working I kept also during my years of study at the university as a for me successful vehicle to become familiar with a certain field of science. Van Wamel once noticed what my method of working was and I could observe from his reaction that he was touched. From that moment on there was some sort of unspoken understanding and appreciation between us.

Physics was taught by Iddekinge. This man was a "type". I will never forget how he demonstrated to us the Doppler effect. The class had to gather in the back of the classroom. Iddekinge stood at the opposite site with a long hollow rubber hose that had a whistle at its end. He started swinging the rope, which had a length of perhaps 3–4 m, at high rotating speed above his head, so that the whistle at the end made large circles, covering a substantial part of the classroom area, and he started to toot with all his might. It was a hilarious scene, but it imprinted the Doppler effect in our minds forever.[5] Once Iddekinge went early in the morning, before the lessons started, to the laboratory of "Hoogovens", the blast-furnace factory in Ijmuiden, at the coast north of Haarlem, where the Noordzeekanaal (= North Sea Canal; see Chap. 2) ends in the North Sea, to fill a Dewar (insulating/thermo) flask with liquid nitrogen. Thereafter, after having transported the Dewar flask to the Lorentzlyceum, he demonstrated in the class a number of effects of very low temperature (liquid nitrogen has a boiling temperature of about -196 °C $= 77$ K) as, for example, the thereby induced brittleness of rubber, etc. Just another effort to teach us physics, or better, as I would now say, materials science, in a way that you could not forget.

---

[5] The Doppler effect involves that the frequency of a wave (here a sound wave) produced by an object becomes larger upon approaching the observer and the frequency becomes smaller upon moving away from the observer. In this case, when the rotating whistle approached the class standing in the back of the classroom the pitch of the whistle became higher and the pitch of the whistle became lower when the whistle moved away.

In the first year of being taught physics at school a second and, as far as I remember, last time I could discuss with my father some basic physics (cf. Chap. 2). My father appeared to be familiar with the experiment with ball and ring according to 's Gravesande (a Dutch scientist from the beginning of the eighteenth century), demonstrating thermal expansion such that after heating a ball it no longer can pass through a ring as was the case before the heating. My father also knew of the experiment according to von Guericke with the Magdeburg hemispheres from the middle of the seventeenth century, exposing the influence of atmospheric pressure, where two hemispheres put together were evacuated (more or less) with the result that the atmospheric pressure exerted on the outside of the combined hemispheres keeps them together and in case of the original experiment, with hemispheres of diameter 50 cm, application of brute force exerted by 16 horses then did not suffice to separate the hemispheres. When my father was a schoolboy he was only offered a limited education, as I mentioned in Chap. 1, but he apparently had absorbed very well what he had learned.

More than 30 years separated father and son, but apparently textbooks on physics had not changed much: they still used classroom demonstrations (ball and ring according to 's Gravesande) and/or descriptions (Magdeburg hemispheres) of these "antique" experiments. Although there is nothing wrong with illustrations of this kind to demonstrate physical principles, it is yet somewhat amazing to observe that even in textbooks of present-day (i.e. after an additional time lapse of almost 60 years) these experiments are still detailed. But no harm is thus done. This becomes rather different if unacceptable "explanatory" text in nowadays textbooks is published. For example, in a current secondary school physics textbook the following text to explain thermal expansion (the experiment with ball and ring of 's Gravesande) is presented (translated into English by me):

"According to the theory of molecules, molecules move faster at higher temperature. In a solid the molecules move back and forth but remain more or less on their position. If the temperature rises, they move faster back and forth and to this end they need more space with respect to their neighbours."

One must have pity with the pupils who have to learn this. First of all, how can it be understood that moving faster to and fro by itself implies that more space is needed? That is baloney of course. Further, the material used in the ball and ring of 's Gravesande is metallic. There are no molecules which, as entities, vibrate: the constituents of metals are atoms and these are the individually vibrating species. Thermal expansion is the consequence of the *short-range* nature of the repulsive force and the *long-range* nature of the attractive force, both acting between atoms.[6] Now one may likely find that the more correct, still simple explanation as presented in the footnote is too complicated for a pupil at this stage to comprehend. However, teaching pupils blatant nonsense is unforgiveable. Then one better presents thermal expansion as a phenomenon only and promises to present later (in a higher/concluding year) a sound explanation or finds a way to "translate" in simpler words the text in the footnote somehow. The above cited text reveals that the author is teaching something he/she does not understand.

Not only were my natural science teachers remarkable. I well remember my Dutch language teacher, (Fries) de Vries. An enthusiastic teacher and a colourful, idealistic, provocative man, who did not hide his strongly pacifist attitude to life during his lessons. He was an active member of the PSP, the "pacifist socialist party". This party was popular and represented in parliament. The party had a young followership. Then and thereafter I did not agree with their main line of thought, i.e. abolition of an army; withdrawal from NATO, etc.; I am realistic about human nature… Years later I found out that de Vries was a poet as well; he never had spoken about that. He published his poems under pseudonym.

(Hein) Wiedijk, also open about his support for the PSP, was responsible for the history lessons. His enormous passion for history and his interpretation of historic events, made the hours with him as something to look out for. Years later, after many years of personal study, Wiedijk published a dissertation and thus got a Ph.D. degree. This fitted well in a tradition: at least in The Netherlands in former days it was not very uncommon to acquire a Ph.D. based on research performed next to a job as teacher. (Van der Waals is the classical example: his groundbreaking thesis of 1873, presenting the famous Van der Waals equation of state for gases and liquids,[7] was prepared while

---

[6] The short-range nature of the repulsive force between a pair of atoms and the long-range nature of the attractive force between a pair of atoms make the curve of *potential* energy of the pair of atoms as function of the interatomic distance, a potential energy well, *asymmetric*. The (maximal) thermal, *kinetic*, vibration energy increases with temperature. As a consequence of the widening potential energy well at higher total (potential + kinetic) energy, the amplitude of the vibration increases (during a vibration cycle the kinetic and the potential energy vary while the total energy remains constant). Simultaneously, with increasing temperature, the *average* bonding distance of the pair of atoms increases as a direct consequence of the asymmetric shape of the potential energy well. Further, in case of a hypothetical, *symmetric* potential energy well, and although the atoms with increasing temperature have higher (maximal) vibration energy (vibrate with larger amplitude), thermal expansion does not occur yet, in flagrant contrast with what one would expect on the basis of the "explanation" from the criticized textbook!

[7] Maxwell (namesake of the not less famous Maxwell equations) wrote a review of the thesis and, as the thesis was written in Dutch, apparently remarked in a lecture that he had learned Dutch in order to study it…

he was teacher at the HBS). Further, especially, say, before 1970, a thesis not very seldom was "a life's work". Wiedijk's thesis is a late example of this.

During the years I followed the HBS-b at the Lorentzlyceum, next to extensive training in Dutch, there was the national obligation to learn three modern languages: French, English and German. No doubt that this acquisition of some command of these foreign languages has led to the impression abroad that Dutchmen were multilingual, "spoke their languages". Unfortunately, not long after I had left the Lorentzlyceum this national obligation to learn these three foreign languages was relieved, as part of the next school reform, and pupils were only obliged to select one foreign language. Unsurprisingly, practically everybody then chose English. The reputation of polyglottism no longer applies in general to the younger Dutchmen. Although I myself do not speak flawless German, even after now having lived many years in Germany, it makes my toes curl to hear some Dutchmen speak German, especially if accompanied by the implicit arrogance that it suffices to use Dutch words, and even Dutch sentence constructions, with a superimposed pronunciation of which the speaker thinks resulting in more or less correct German. This happens under the misleading supposition that German and Dutch being related allows such approach.

One final anecdote must be recalled here. One biology teacher, Muller, who only taught us one year, had a remarkable way of evaluating the test-papers of his pupils. If in your answer to a question/problem posed (i) you had written something both correct and fitting to the question/problem, you got a positive point. If in your answer (ii) you wrote something correct but *not* fitting to the question/problem, you got a negative point; if (iii) you had written something incorrect, than obviously you also got a negative point. Eventually, for the entire test-paper, all positive points were counted and from this number all negative points were subtracted (i.e. Σ (i) $-$ {Σ (ii) $+ \Sigma$ (iii)}). You could thus end up with amply more than 100 points per test-paper. Muller had adopted this procedure as he felt that in this way a fairer evaluation of the pupil was possible: Indeed, assigning a negative point to a correct but for the answer to the question irrelevant statement introduces an element in the evaluation of the quality of a response that is usually not considered; it forces the pupils more carefully than usually to consider what they incorporate in their responses, so that these become more focussed. Reviewing all test-papers of the class on this basis was an enormous task. Predictably, during the year not many test-papers were made. In all my years as Professor at a university I have never observed that anybody else had adopted this painstaking approach of evaluation of test-papers/exams, neither have I. But I have sympathy for the method of evaluation of my biology teacher.

During these years Eric continued studying intensively piano playing. He appeared to be pretty good in it. His piano teacher, Cor Rackwitz, wished him to attend a college of music (called "conservatorium" in The Netherlands) after having finished his education at the Lorentzlyceum. Eric decided against it and has never regretted that: the chance to become a concert pianist of some standing, playing music composed by others, is very small and Eric envisaged greater possibilities in other, as he assumed to be, intellectually more satisfying, directions. But piano playing, also making music together with others, accompanied him during his whole life, although long periods could pass with lack of time and energy to practice... It always served and still serves as the perfect tool for relaxation.

To Eric's great distress his piano teacher Cor Rackwitz suddenly became ill and died. Eric, accompanied by his mother, went to the funeral home where the deceased was, so to speak, laid out. It was the first time Eric saw, vis-à-vis, a dead man, a corpse. It was an unpleasant sight. Cor Rackwitz's face was difficult to recognize as it deviated strongly from how it had looked like alive; the already in living state small body seemed to have shrunk considerably. The colour of the skin was strangely yellow. This image haunted Eric for considerable time in his dreams thereafter.

During my HBS-b years I gradually developed a vision of my future (that soon would prove to be rather incorrect). My initial intention was to become an engineer. This wish was partly derived from a somewhat romantic idea about what an engineer's role in society could be. And engineers were held in high regard. I had read a booklet with the title "Ingenieursdromen van de toekomst" (= "Engineer's dreams of the future"), which had pronounced impact on me. Most projects discussed in this booklet would require input from in the first place *mechanical* engineers. However, during my first HBS-b years physics and chemistry had a strong appeal on me, i.e. I became more science oriented, although that was not very clear to me at that stage. Thus I decided that I would not become a mechanical engineer, and, after ample consideration, I believed that "chemical technology" would be more appropriate for me. This is perhaps rather surprising, as in the above text neither I have written about my experiences with chemistry during my HBS-b years, nor did I mention my chemistry teacher, Mulder. Mulder was a rather young man having finished his academic training shortly before. He certainly had a passion for chemistry that he could convey to his audience. Many years later he, to my surprise as I hadn't have contact with him after my time at the Lorentzlyceum, attended my Inaugural Lecture as Professor at the Delft

University of Technology in 1985, which I much appreciated. Also, during my school years I enthusiastically performed many chemistry "experiments" in the barn in the garden of my parent's house (yes, "explosions" were part of it). So looking back, at the time I must have been a chemistry "adept". Hence I chose chemistry as my field and saw myself in the future as an engineer looking after oil fields and associated plants, e.g. in the Far East. The reserve that I nowadays have towards pure, especially synthetic chemistry, as the reader will be made clear later in this booklet (see Chap. 5), is likely the consequence of a professional "turn" I made during my first Delft years.

In The Netherlands there are three universities of technology; the most renowned one is the oldest one: the Delft University of Technology. The academic degree to be acquired after five years (i.e. if the formal schedule is followed; most students needed a couple of years more) is "Ir.", an abbreviation of "ingenieur" (for example my mathematics teacher, van Wamel (see above), carried this degree). Similar degrees in Germanic countries are indicated as "Dipl.-Ing.". As far as I know not many other countries had such engineer degrees (Poland had). Nowadays the university educations in European countries have adopted the Anglo-Saxon model. Thus the "Ir." Degree is replaced by "M.Eng." (master of engineering; general university degrees at the same level are indicated as "M.Sc." (master of science)). But these degree-name changes happened long after my time as a student. So in the end I opted for the Delft University of Technology, where I would study chemical technology and would become a chemical engineer ("Ir.").

Epilogue: The Downturn of the Lorentzlyceum

In the following lines the rise and decline of the Lorentzlyceum is briefly sketched. Of course a large part of this development has not at all been part of my life (I was a pupil at the school from 1962 till 1967). However, in this way unhappy changes which have led to the disappearance of what can be rightfully called an elite school community are illustrated, and thereby stand model for more or less similar developments elsewhere in the same period of time.

The Lorentzlyceum started as a pure HBS-b in September 1919. Already in its first decade the number of pupils increased remarkably, reflecting that this school type gained ever more followers (see the beginning of this chapter). The city council of Haarlem responded to this development and, after having been housed in what can be called temporary accommodations, in 1929 the school moved into a nice, one could even say beautiful, new building at the Santpoorterplein (see begin of this chapter). Now the school became a lyceum, implying that a gymnasium department was included. This is the school and building I spoke about above (Fig. 3.1) and where I spent my HBS-b years much later.

The school flourished; whereas in the middle of the fifties about 300 pupils attended the Lorentzlyceum, in the middle of the sixties (I finished my school years in 1967) about 650 pupils followed there either the HBS-b, HBS-a, gymnasium-b or gymnasium-a directions (where the "b"-variants imply the educations with emphasis on mathematics, physics and chemistry). As the culmination point of the development and growth of the Lorentzlyceum one may see the move in 1971, with about 700 pupils then, to a new building at the Sportweg/Planetenlaan in the northern part of Haarlem. This building in no way has the same level of architectural appeal as the former building at the Santpoorterplein; it simply is an example of unimaginative, utility architecture.

Around 1980, likely as a result of the demographical development, involving reduction of the number of births per married (or unmarried) couple, the number of pupils had decreased. This led to closure of the gymnasium department. But this was only a side effect. The drop in the number of pupils became such distinct that the city council of Haarlem forced a fusion of the Lorentzlyceum with the Klaas de Vries school for MAVO. The MAVO was the more or less renamed MULO, a secondary school of lower intellectual rank than the HBS or gymnasium, and thus lyceum. No wonder that the "Lorentzlyceum" was strongly against such fusion, but their protests were to no avail. Thus in 1984 the unwelcome fusion was imposed. This was the end of the Lorentzlyceum.

A number of problems emerged as a result of this unhappy, politically enforced change.

The first debated issue was a new name for the fused school. In the end not even the name "Lorentz" could be kept: the proposal "Lorentz Scholengemeenschap" (= "Lorentz school collective") was not accepted. In the end the uninspiring, unpretentious name "Schoter Lyceum" was agreed upon (Schoten was the name of a former village at the location of what was now Haarlem-Noord (Haarlem-North)). However, upon loud protest of a gymnasium in Haarlem, the word "Lyceum" had to be dropped as a gymnasium department was no longer incorporated in the school. Finally the name "Schoter Scholengemeenschap" (= "Schoter school collective") crystallized.

Obviously, joining teachers of such different intellectual level in one team (the teachers of a lyceum were normally academics (see the beginning of this chapter), which does not hold for those teaching at a MAVO), unavoidably is associated with differences in status (and salary) and this may hinder the development of harmony and unanimity among the staff of the school resulting from fusion of in this sense unequal partners. The difficulties became such severe that they were even subject of a substantial article in a national newspaper.

Of course, those associated with the former Lorentzlyceum feared loss of reputation of the school. Indeed the standing of the Lorentzlyceum could not be maintained by the Schoter scholengemeenschap. Voices were heard that on a school solely for the highest form of secondary education (such as the still existing "stedelijk gymnasium" in Haarlem (and the former Lorentzlyceum)) pupils in the same year were more advanced than pupils following the in principle similar education on the Schoter Scholengemeenschap.

In "school collectives", pupils in their first year at school, were put in first-year classes called "brugklassen", where those in principle qualified for a higher school form were combined with those for whom a lower school form was more appropriate. This approach was partly meant as a justified remedy to detect "late developers" under the pupils in this first year at a secondary school. The very much stronger, political motivation behind was to provide access to higher education to an as large as possible part of the population, which can only be considered as a positive strategy. The damaging side effect, however, was the inflow of unqualified pupils in the highest/higher form of secondary school, which of course reflected the result of pushing by parents; admission examinations, as I was subjected to in 1962, had been abolished…

It has been argued before (see *"Damage a School Can Cause"* in Chap. 2) that joining pupils of significantly different intellectual powers is perhaps an advantage for the lesser qualified pupils, as these may feel stimulated, but the best pupils in this system are underchallenged and paid less attention to, which brings about pronounced societal loss; i.e. waist of talent.

The Netherlands is a country of strongly egalitarian nature. During the times considered here the debate "nature versus nurture" was strongly inclined to "nurture", i.e. the tendency was to assume that more or less "everybody" could be elevated to a state of higher/high development provided the education and the social circumstances/economic boundary conditions were optimized to this end.[8] Only much later, at the time of writing this text, it

---

[8] To illustrate the "mood" in The Netherlands in these days, I may digress here and refer to the fate of Professor Wouter Buikhuisen. Buikhuisen was appointed as Professor of Criminology at the University of Leiden in 1978 and devoted his research to the relation between criminal behaviour and biology. He was of the opinion that criminal behaviour can (also) be determined by biological factors, i.e. can be "innate". In other words: "nature" could dominate "nurture"; an *inborn* predisposition for criminality can be prevalent. Thus Buikhuisen wanted to investigate the relationship between criminal behaviour and brain functions. This intention and especially his ideas raised great uproar amongst many, especially leftist "intellectuals" in The Netherlands. Buikhuisen was compared to the nazi-physician Mengele and even personally endangered. In journals and weeklies a concerted action against him and his ideas was made. An esteemed journalist/author, Hugo Brandt Corstius, took

has become more accepted that talent, and thus high intellectual power, is to large extent "inbred", i.e. "nature" overrides the possibilities of "nurture".

As I have learned from a lifelong experience with learning and studying myself and teaching students, there is no need to reform a working learn/study method as long as a new one is not proven to be better. In fact this is a serious problem in the education system presented by primary and secondary schools: too often a teaching method is changed without a significant scientific basis validating the imposed method change, i.e. demonstrating better

---

the lead in the damnation of Buikhuisen. As a public figure Buikhuisen was "killed". The perhaps most troubling and alarming consequence was the lack of loyalty of his employer, the University of Leiden: Buikhuisen was transferred to another department and his financial means for research were minimized. It is highly unscientific to not let follow a reasonable line of research in a certain field of science just because of deviating public opinion; it is in drastic conflict with the sacred academic freedom and a shameful chapter in the history of the University of Leiden. In 1988 Buikhuisen completely withdrew from science and started a living as a seller of antiques… Many years later the taboo on research on finding the biological factors giving rise to, for example, criminal behaviour was released. The University of Leiden more or less apologized. Buikhuisen was invited to give a lecture at the university. He followed the invitation and in April 2010, for a large audience, he delivered this lecture which was met with positive response.

This whole episode, that I had followed closely (I had a subscription for the weekly that published the columns of Brandt Corstius and I left The Netherlands not before 1997/1998), illustrates how important academic freedom is; it should be a sacred good (cf. Chap. 7). And more generally: we now must have learned that people having the same rights does not involve that they are equal in their possibilities, or, in other words, "nature" implies that congenital quality differences between human beings are unavoidable. This limits the role of "nurture". And, to return to the message of the main text above, this limits the sense of school collectives and their "brugklasses" as well.

Academic freedom is not a matter of self-evidence. Thus, at least some, also academic audiences are not prepared to let express opinions which are in conflict with their own, moral or political convictions or beliefs. The recent clash on "gender identity" serves as an example. At a number of universities, especially in the English speaking world, massive numbers of students, and also of the academic personnel, have raised protests against lecturers advocating the dominance of biological characteristics to define if someone is a man or a woman. Instead, the protesters claim that how someone "feels" defines that person as either a woman or a man. The loud protest of these "trans activists", accentuated by setting up, but not limited to, blockage of lecture halls, has led some university boards to uninviting originally sought for lecturers (usually from outside the university concerned), who are expected to be representatives of a critical attitude towards "gender ideology". This can go that far that some academics expressing opinions critical of "gender ideology" have experienced negative career consequences. It is worthwhile to remark here that the conflict, as a matter of fact, is only about how to define and distinguish between man and woman; it is, of course, left free to anybody to "feel" as a man or a woman, or as "LGBTQIA + " (= lesbian, gay, bisexual, transgender, queer, intersexual, asexual or any other unconventional sexual orientation not listed here); i.e. an equivalency of all these sexual orientations is not impeded by a discussion about how a man and woman are to be defined. The imposition of a public opinion, implying that "feelings" are claimed to override biological facts, represents a case that bears an intrinsic parallel with the Buikhuisen controversy discussed above. Thereby the merit of academic freedom, also involving debate particularly between adversaries, is severely damaged. This smothering of discussion, in fact suppression of opinion, is well known for dictatorially governed countries. But, remarkably, the case discussed concerns genuinely democratic countries as well: a deviating public opinion may abort the possibility for letting someone, presenting an adverse, "unwelcome" point of view, be heard.

results: do pupils nowadays learn faster and/or better to read, write and calculus? This leaves unimpeded that tools are developing that we should use: computers are likely the most obvious and drastic example of recent days. But this does not affect any essential mechanism of the learning/studying process.

4

The Notions Science and Physical Law

Abstract A defining description of science is offered. Roles of theory, models and the experiment. The physical law as representing the order inherent to nature. Can we know it all?

Science is the activity of mankind to unravel the inner workings of nature. It starts with making observations of specific phenomena on a certain system, say planet movements within the solar system or deformation of a solid body under the action of a mechanical force, etc., etc. Then an intellectual process can start to explain the observations in terms of some model of the relevant part of nature, i.e. the system. Such a model is not the reality. It is simply the result of our theorizing to provide a unifying concept for a certain set of observations for which "understanding" lacked. One can only speak of a viable model if, in a next step, the model allows correct predictions for the system concerned under conditions different from those pertaining to the observations leading to the development of the model. Hence, as long as these predictions are in agreement with the results of the corresponding experiments it is said that, *for the time being*, the model provides a satisfactory description of nature.

Progress in our understanding of nature is initiated as soon as the observations are no longer in agreement with the currently accepted model, i.e. deviations from the predictions are measured. This immediately makes clear that making ever more accurate measurements of even the same phenomenon is a prerequisite to arrive at more profound understanding. For example,

E. J. Mittemeijer, *How Science Runs*,
https://doi.org/10.1007/978-3-030-90095-3_4

referring to the above examples of systems: (i) an initial model for the revolution of the planets around the sun involved that the orbits of the planets were circular. More accurate measurements revealed that the orbits are better described as ellipses with the sun in one of the foci. This eventually led to Newton's law of gravitation[1]; (ii) the reversible, elastic deformation of a body upon the action of a mechanical force is described by Hooke's law.[2] This elastic deformation was initially thought to be independent of the direction in which the force acted. More accurate measurements revealed that this is not generally the case and that this direction dependence, "anisotropy", is closely related to the arrangement (and bonding) of the atoms in the body. This eventually led to the generalized Hooke's law.

The above text suggests that the primate in science is reserved for the experiment/the measurement. As I have written elsewhere: "The result of theorizing is uncertain and most subject to decay: *observations are immutable, but insight can become deeper and more comprehensive in the course of time.*" (further see: *E.J. Mittemeijer, Fundamentals of Materials Science, Second Edition, Springer, 2021, pp. 6–10*).

The above text also makes clear that science does not deliver absolute truths. We may say that we understand ever more of nature, but that the border of our knowledge moves away from us: we appear to be unable "to grasp it all". Or, more hopeful, it could be said that we approach full understanding of nature asymptotically. Yet, science does allow us to make reasonable, educated guesses about the unknown, and that is also science in action! This means that we cannot prove such educated guess, but we can justifiably say that a specific phenomenon is highly likely or unlikely. Thus, as an example of an educated guess: finding human life *like us* in our galaxy appears unlikely, as far too many rare, unusual concurrences have to occur (this is the reasoning according to *J. Gribbin in: Scientific American, September 2018, 86–91*), but we cannot prove its impossibility.

How then does nature disclose itself to us? In a direct way this is realized by the physical laws (the adjective "physical" is used to contrast with

---

[1] The law of gravitation expresses that the gravitational force, F, acting (along the line) between two bodies, A and B, taken as point masses, is proportional with the product of their masses, m_A and m_B, and the reciprocal of the square of the distance between them, r_{AB}: $F = G.\frac{m_A m_B}{r_{AB}^2}$, with G as the gravitational constant.

[2] Hooke's law expresses that the mechanical force, F, acting on a body introduces a change of length, Δl, in the direction of the force, which is proportional to the force: $F = C.\Delta l$. The constant C is called the stiffness of the body concerned. In order to avoid the trivial effect of size for bodies of similar shape and of the same material, the force is normalized to stress, σ (= force per unit area), and the length change is normalized to strain, ε (= relative length change). Then the usual notation for Hooke's law for elastic deformation is obtained: $\sigma = E.\varepsilon$, where the constant E now is a (size independent) material constant called the elastic modulus or Young's modulus.

"political" laws). Two of these have been mentioned above. Physical laws are found on the basis of empirical work (i.e. relying on observations). The physical laws can be conceived as expressing the underlying "regularity" of nature. The physical laws cannot be derived; they must be considered as "translations (into mathematical formulas)" of sets of observations. In other words: we cannot prove them….; they are the qualities of nature.

Confusingly sometimes a relation that can be derived in a straightforward manner from existing viable concepts/theory is also called a "law", as Fick's second law (see Sect. 9.2) or Bragg's law (see footnote 8 in Chap. 5), which "laws" were indeed *initially* derived theoretically in a straightforward manner. Evidently, within the context of the present discussion, these relations are less expressive of the inner workings of nature than the genuine laws indicated above.

Upon progress of science insight may be acquired that allows to find out that for a genuine physical law, indeed once identified on the basis of empiricism alone, it can be possible to present a theoretical derivation on the basis of "deeper lying" "regularities" of nature. In the above sense such a law then cannot be considered as a genuine one anymore. In practice we retain the notion "law" also for these cases. This holds for the second example dealt with here: Hooke's law. This law was initially identified on the basis of observations alone. However, later it could be shown, adopting atoms as the constituents of matter, that the background of this law and thereby its derivation derives from the potential energy describing the bonding between (a pair of) atoms as function of the interatomic distance.

Apparently, this last discussion leads to the visualization of science as proceeding by peeling off skin after skin of an onion, representing nature, where, to picture the endlessness of this process, these skins become ever thinner. The maintenance of something unknown may be seen as a principal property of nature. Or, after all, in the end of physical research complete understanding of nature is achieved yet...

5

The Becoming of a Scientist

Abstract Study of chemical technology at the Delft University of Technology. Change to a more science rather than engineering oriented study. The role of outsiders in science. How totally different ways of lecturing by, characterally, totally different professors can be very fruitful for a becoming scientist. A late experience of what is learned during a study, subsequently "forgotten", suddenly "awakes", decades later, at the end of a career. Turn to solid state chemistry/physics. First involvement with (X-ray and electron) diffraction research at the Laboratory of Metallurgy during the Master project, the departing point for the following Ph.D. project. First experiences at international conferences. Career considerations; the Philips Research Laboratories versus the University. A reflection on the development of careers from science to management.

On a dry, but cloudy day in late summer 1967 Eric arrived by train from Haarlem in Delft for finalizing his enrollment at the Delft University of Technology (= Technische Universiteit Delft, TUD). Eric had never been before in Delft and had to find his way using a map (it was long before the time of smart phones with navigation apps…) going by feet from the railway station, crossing the "Rijn-Schiekanaal" (kanaal = canal) via the at that time still existing Rotterdamse Poortbrug (brug = bridge), to the administrative/main building at the "Julianalaan 134". He was a bit excited.

Delft is a small city with a historical centre that attracts many tourists. Its atmosphere has a distinct accent imposed by the students and personnel of the university, as also experienced in other university cities of modest size and with century-old centres, like, for example, Tübingen in Germany. The TUD is the renown, oldest university of technology in The Netherlands. It has a history originating in the middle of the nineteenth century, but, via an intermediate stage as a polytechnic school, its recognition as a university was realized only in 1905. In former days the university buildings were largely

E. J. Mittemeijer, *How Science Runs*,
https://doi.org/10.1007/978-3-030-90095-3_5

situated in the old centre of the city. In 1967 the university buildings were concentrated in a separate quarter, located outside and south from the centre, i.e. south from the "Rijn-Schiekanaal" and at the eastern side of the canal "Schie", extending to the eastern edge of the city.

After having completed the admission formalities. Eric was informed that he could immediately start the following day with a practical course on anorganic chemistry. This was a message he had to digest for a moment, as he had expected and was internally set to that the actual study would begin a few weeks later with the start of the academic year. Thus an abrupt change to his plans would have to occur. The department chemical technology offered this possibility in view of the experience of many preceding years that it would be rather difficult for a student to have completed all lecture courses and practical courses of the first year (indeed) at the end of the first study year. So instead of a few weeks more holidays, Eric did start the following day with preparing and performing his first "real experiments" in the special, separate building for first year chemistry students ("Gebouw voor Scheikundige Propedeuse") at the "de Vries van Heystplantsoen" ("plantsoen" = park), a building situated close to and more or less opposite to the above indicated main/administrative building of the university.

This practical course anorganic chemistry only involved synthetic work: on the basis of a description in the literature, one had to construct laboratory set-ups allowing the synthesis of a few compounds. This first confrontation with his study had a rather negative effect on Eric and that is the reason to dwell upon it here.

Eric had the feeling that this type of student training was very old fashioned. There was no science offered: the students more or less were like cooks executing recipes according to the (rather old) literature they had to read first. Insight, e.g. why nature allowed these reactions to occur and, equally important, others not, was not acquired at all. But of course, a lot of techniques, associated with the construction of (glass) set-ups for chemical reactions to run, were learned in this way. One may state that this is an important aspect of especially chemical synthesis research, and genuine laboratory experience is a prerequisite for experimental scientific work in general. But it is doubtful if it is wise to begin a chemistry study in this dated way. Eric began to lose interest in chemical synthesis. Yet he finished this practical course, in which, admittedly and perhaps understandably, he had not been very successful. Luckily his performance sufficed to avoid repetition of this course. Eric was happy that the lectures now began…

This early bad experience invoked uncertainty in Eric. He was no longer fully sure to have the talents allowing a successful study at the TUD. This worry was enhanced by the realization that of the about 165 chemistry students who had started the study in 1967, about 50% would have left the university in a year or so.[1] As a response Eric was very much determined to prove to himself that he could do it.

In the following some essential experiences Eric made during his study years in Delft, are reported, in as far as they both illustrate how they influenced distinctly his development and are of interest to a reader of this booklet.

Many studies in the natural sciences have in their first year one more or less general lecture course that must give the freshman an impression, perhaps even some sort of overview, of the field of science selected. This is the more important as in the first one or two years of a study a lot of courses in other fields have to be followed (for chemistry students: especially courses in mathematics and physics). Such first year course on "general chemistry" was delivered by G. W. van Oosterhout. For Eric this course was a relief, after having gone through the above sketched practical course. Although many topics were dealt with, and mostly only in a rather superficial way, in all cases van Oosterhout was capable to open small doors to even, if only little, deeper insight into the phenomena discussed. Regrettably his lectures were rather chaotic, not well structured, but in combination with self-study one had the feeling to move "forward". The rambling character of van Oosterhout's lectures will certainly have led to this lecture course developing into a major barrier for many of the students.

Looking at the Periodic System; The Role of "Outsiders" in Science

An essential aspect of science is to be able to view a problem or phenomenon from various angles. In the general chemistry lecture course by van Oosterhout this was illustrated, for a very simple case, by showing more than one reasonable way to present the Periodic System/Table.

---

[1] In his welcoming speech to the first-year students the president of the university, called "Rector" in The Netherlands and Germany, had the, I suppose, custom to say that every person in the audience should look at his/her neighbour and say to him/herself: either you or I are no longer there in the next year. I am not sure if that menacing statement was a stimulating remark for everybody in the audience…

The usual way of presenting the Periodic System is shown in Fig. 5.1a. The elements are arranged in a sequence of increasing atomic number, Z (=number of electrons in the (neutral) atom = the number of protons in the nucleus). Starting at the top of the Table, from left to right in the first row the atomic number increases: $Z = 1$ and $Z = 2$. In the next row, at one level below the first row, the next elements are also arranged according to increasing atomic number: from $Z = 3$ till $Z = 10$; etc. The rows in the Table thus constituted are called "periods"; the columns are called "groups". The sense of this presentation, as originally devised by especially Meyer (1864, 1868) and Mendeleev (1869, 1870, 1871), is that the elements in a group have similar (chemical, physical) properties. This presentation thus depicts pure phenomenology; initially no understanding existed for this periodicity in properties. Therefore, with reference to the discussion in Chap. 4, it was justified to name this periodic phenomenon "The Periodic Law", as there was no way to derive this law "in a straightforward manner from existing viable concepts/theory". Much later, as the result of quantum mechanical theory, an explanation for this periodicity was presented: the arrangement of elements in a group is due to similar electron configurations in the "outer" shell of the atoms in the group. These "outer" electrons are the electrons of highest energy, are relatively weakly bound and have a distinct probability to be at locations farthest away from the nucleus and therefore they play an important role in establishing bonds with other atoms, i.e. they determine the reactivity of the element concerned.

To expose the role of the electron configuration for the Periodic Law, a presentation of the Periodic Table can be made such that the electron configuration of the atom directly emerges as the guiding principle for the Periodic Law. Thus the "s-block", "p-block", "d-block", "f-block",…. develop, as shown in Fig. 5.1b. Because of its morphology this variant of the Periodic Table is also called the "left-step Periodic Table". This arrangement of the elements can be seen as physically more revealing than the traditional presentation shown in Fig. 5.1a.

Fig. 5.1 a The Periodic Table, in its usual presentation. The elements in a vertical column (a "group") have similar properties (taken from "E. J. Mittemeijer, *Fundamentals of Materials Science*", Springer, 2021). b The Periodic Table, according to Janet and Dockx, "left-step Periodic Table". The elements are presented in a way emphasizing the electron configuration of the atom as guiding principle for the Periodic Law (taken from E. J. Mittemeijer, "*Fundamentals of Materials Science*", Springer, 2021; Sect. 2.5).

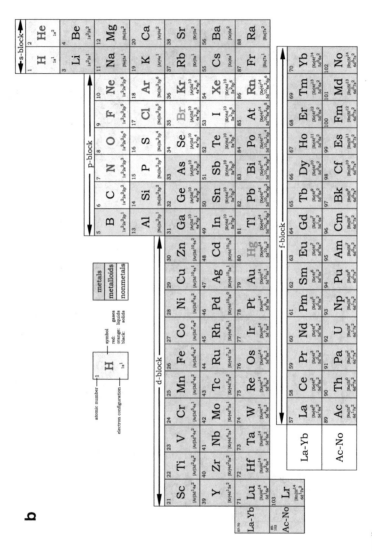

Fig. 5.1 (continued)

A striking feature of the version of the Periodic Table shown in Fig. 5.1b is the appearance of helium (He) at the top of the column with the alkaline earth elements (i.e. on top of beryllium (Be; row 2))). This reflects that He has a filled (outer) *s* orbital (i.e. with 2 electrons), as also holds for the alkaline earth elements. However the inert character of He seems to reserve for He the place on top of the column of noble gases......; at least that could be the preference of chemists.[2] This may be the reason that the Periodic Table as presented in Fig. 5.1a is still the more popular one. Yet, evidently, the presentation in Fig. 5.1b directly reflects the electron configuration as guiding principle for the constitution of the Periodic Table. This must have been the reason for emphasizing this approach by van Oosterhout.[3]

In the lecture hall in Delft, where van Oosterhout gave his lectures, a huge version of the Periodic Table as presented here in Fig. 5.1b was permanently hung up on the wall behind and above the lecturer. The originator of this version of the Periodic Table was indicated as S. Dockx. The name meant nothing to me for a very long time. Only very many years later I bothered to find out who S. Dockx was, as I had noticed that outside "Delft" no colleagues of mine apparently had ever heard of him as the originator of Fig. 5.1b.

Stanislaw I. Dockx (1901–1985) was a Belgian monk ("Father Dockx"), who published many religious works and, rather strikingly, one scientific work, a book called "*Théorie fondamentale du système periodic des éléments*" (= A fundamental theory for the periodic system of the elements). This book, published in 1959, which takes a singular position within Dockx's many theological, religious publications, originated from work done much earlier, likely in the years from 1929 till 1936, while he took courses on physics and mathematics at universities (he never acquired an academic degree in natural sciences; he became a doctor of theology in 1938). His book on the Periodic System is written in French, which certainly has been an obstacle for

---

[2] Yet He can form stable compounds, as a He-Na compound, Na_2He, that is stable at high pressure (>115 GPa) (*Xiao Yong et al., Nature Chemistry 9 (2017), 440*). It was shown earlier that other noble gases could form stable compounds.

[3] At this place it is appropriate to remark that, whereas the position of He in the version of the Periodic Table, that may have the largest appeal to physicists (Fig. 5.1b), may be seen as "critical", also in the traditional version of the Periodic Table, that may have the largest appeal to chemists (Fig. 5.1a), uncertainty about the position of one element can be indicated: hydrogen, H, positioned in Fig. 5.1a on top of the column of the alkali elements (i.e. on top of lithium, Li), may also be positioned on top of the column with the halogens (i.e. on top of fluorine, F), as it has both the tendency to give away its *s* electron (as the alkali elements) and the tendency to fill up its (1*s* and naturally) outermost shell with one additional electron (as the halogens).

worldwide dissemination, apart from Dockx being an outsider in the scientific world. In fact it appears that the proposal by Dockx was preceded by one of more proposals due to Charles Janet.

Charles Janet (1849–1932), a Frenchman, was educated as an engineer. Thereafter he became a doctor of natural sciences at the Sorbonne. After his marriage he became a successful manager of the brush factory owned by his father-in-law. Next to his daily job as manager of the brush factory, he was a productive inventor and an ardent private researcher in a number of branches of science, especially entomology, biology, geology and paleontology. At the advanced age of about 78 (i.e. in 1927), he started his work on the Periodic System. He presented his proposals in a series of privately published papers, which were written in French. At the time these papers remained practically unnoticed. Apparently also in Delft in 1967 one was unaware of Janet's work. Only in much more recent years the work by Janet was worldwide rediscovered.

The above historical sketch demonstrates a nowadays rare, distinct contribution of "outsiders" to physics/chemistry. Currently, amateurs or hobby-scientists (the latter qualification certainly applies to both Dockx and Janet, and is meant respectful) can still contribute to the progress of science, but this is nowadays apparently restricted to fields such as paleontology and astronomy, by finds and observations, respectively. Acquisition of fundamental insight, however, appears to be reserved to professional scientists.

The above discussed proposal for a Periodic System was the only contribution to science by Dockx. He was a religious "official". It may be speculated that the corresponding (limited?) occupation, and the assured personal situation may have provided him in the years 1929–1936 the possibility to develop his ideas on the Periodic Table in his free time and on his own. It then took many years before he published this work in 1959. His education in the natural sciences was fractional and this may have led to errors in his book which was severely criticized in a review that his book received (*R.F. Trimble Jr., Journal of Chemical Education, 38 (1961), 222*).[4] His contribution to science still remains significant.

Janet resembled a type of scientist of some centuries ago: a "man of means", i.e. a man in a personal, well-off situation with time enough to devote a lot of effort and finances to perform scientific work as a private person. The record (and range) of his scientific works is large: over 100 publications representing more than 4000 pages printed text. His very late work on the Periodic System

---

[4] I am indebted to Eric Scerri (University of California, Los Angeles) for discussion and making me aware of this review.

constitutes only a small fraction of this output. He was a most remarkable man.

In my career quite a few times I have been approached by "amateur-scientists" who presented theories, laid down in extensive manuscripts, on solutions for existing "problems". Yes, demonstration that a "perpetuum mobile" would be possible also belonged to these manuscripts, possibly as a consequence of thermodynamics having been indicated as a field of importance for the research of my Chair/department. The authors of these manuscripts longed for recognition and it could be a laborious job to exactly pinpoint the essential error/flaw of reasoning unavoidably made by them.

Dockx and Janet, although being outsiders in the fields of physics and chemistry and presenting their work in, what can be called, even obscure or at least hardly visible publications, in no way can be considered as such "amateur-scientists". They are both genuine hobby-scientists and mavericks. Their contributions are significant. It would be fitting and more than timely that the Janet-Dockx variant of the Periodic System enters textbooks.

The late recognition for the work by Janet and Dockx makes us aware that an unknown outsider may contribute something essential to the progress of science, and that we are inclined to overlook it.

The beginning chemistry students in Delft had to follow the course analytical chemistry which was given by K. J. Metman. Metman was a person totally different from van Oosterhout. He was always (more than) correctly dressed and behaved accordingly, i.e. with civilized, formal manners. His lectures were very clear, very well structured; on the basis of the notes one made and elaborating them at home later (see what is written in Chap. 3) one could successfully master the concluding examination. That could not be said from the lecture course by van Oosterhout…, where a lot of supplementary self-study using additional literature was a prerequisite. However, the way of studying required for the van Oosterhout lecture course is obviously much better representing and teaching the attitude required for a successful researcher. Such much more time consuming practice *must* be experienced, but one may wonder if that has to happen with the very first lectures of a study. It does not surprise that the Metman lectures were appreciated more by most students.

I have never understood the position of Metman within the faculty. He had the aura of an important professor. Also his, if my memory serves me well, large and impressively furnished office in the first floor of the building at the de Vries van Heyst plantsoen, above the main entrance, radiated this atmosphere. Yet, the academic degree of Metman was only "Ir."; he not even had a doctorate and, as far as I could retrieve now, he had published only rarely.

The most important findings to be gained from a lecture course need not be directly core business for the topic of the course. The type of problems that were dealt with in the analytical chemistry course by Metman often required the solution of a rather large set of simultaneous, cumbersome equations. Metman showed, and Eric learned, that by appropriate omittance of some terms, which at some place in the equation considered are of negligible value with respect to other, specific terms, and checking all equations for application of this approach, even rather accurate solutions can be obtained by then analytically solving the resulting set of equations. Such treatment involves understanding of the physics/boundary conditions behind/pertaining to the problem. Of course, even hand-held computers can nowadays usually numerically provide exact solutions for problems as meant here. But the type of assessment indicated presented an important recognition to Eric, that helped him in his later, own research, as it led to insight into (the physics of) the problem, that is not acquired by blunt application of "brute force" solutions.

Eric was now doing well and, as he got rather good marks in mathematics courses, he considered the possibility of transferring himself to the mathematics faculty after the first year; the TUD offered this possibility such that one could directly continue with the second year of the mathematics study. No doubt his initial experience with chemical synthesis work pushed him a little in this direction. However, Eric stayed with chemistry because, after all, he was much more attracted by obtaining more knowledge on how nature works; mathematics has great beauty but does not directly uncover the acts of nature.

The last time Eric was confronted with chemical synthesis experiments was the obligated practical course on organic chemistry in his second year. He passed successfully, but he now was absolutely firm in his decision to possibly avoid any such activities in his further study.

After the second year the chemistry student had to choose one of four possible directions for his/her further study. The distinctly largest part of the students obviously chose the direction "*chemical technology*", which in any case represents the main motivation for studying chemistry in Delft. Eric had decided for chemistry in Delft for the same reason… In the first two years of study, however, Eric's ideas about this study and future professional life had pronouncedly changed. He had become much more "science" oriented than "engineering" oriented. He no longer saw himself as a future engineer working on a plant somewhere on the world. Therefore he chose the direction "physical chemistry", which in the end promised a more science based career.

Now and then I have been asked, in view of my later career, if it had been a mistake to study in Delft at a university of *technology* rather than on a general university (as in Leiden or Utrecht, or Groningen). And wouldn't it be better for me to have studied physics instead of chemistry? Of course, definitive answers to both questions are not easy to give.

The focus on technology during my study in Delft, and the science based engineering "atmosphere" there, has served me well: in my later research work, that can be characterized indeed as research on fundamental scientific problems, in many cases yet an eye was directed on a technological application that needed deep scientific understanding not available then. I even, but admittedly rarely, contributed to solving an engineering problem. Against this background I still feel at least partly also as an "Ir.", and never considered dropping my membership of the "Royal Netherlands Society of Engineers" (=Koninklijk Instituut Van Ingenieurs, KIVI).

As I later developed into, actually, a materials scientist, I can state that it is rather irrelevant to distinguish between a completed physics study or a completed chemistry study as basic academic education before entering the field of materials science (see my notes on materials science as a discipline in Chap. 8). Moreover, as any scientist in the natural sciences experiences: the borders between disciplines as chemistry and physics do not really occur; they are at best very vague. The delineations of a scientific discipline, which non-scientists and beginning students believe to exist, evaporate upon one more and more delves into science. However, differences of the preceding basic study curricula can be of significance. Thus, in physics studies usually more mathematics is taught than is the case in chemistry studies. After having selected the physical chemistry direction after the second year, I felt the need for and accordingly realized more mathematics training. I have met many chemists then and later who, for example, did not master Fourier and Laplace transformations.

I was glad that, having completed successfully the first year, most time was now spent in the building "Gele Scheikunde" (= (directly translated) "yellow chemistry"; see what follows), at the Julianalaan 136, which was the centre of the faculty (Fig. 5.2a). This building got its name because it was constituted from yellow bricks, which, however, in my time were no longer very yellow and had turned more or less to grey (Fig. 5.2b). I found the building appealing: it had only two stories; its main entrance was inviting; the building was not as imposing as some modern many storey university buildings housing huge faculties. As a consequence "Gele Scheikunde" was

a

b

Fig. 5.2 The main building of the Faculty of Chemical Technology: "Gebouw voor Scheikunde" ("Gele Scheikunde"; the Dutch word for "chemistry" is "scheikunde") at the Julianalaan 136; used from 1946 till 2016. **a** A photograph, from shortly after the Second World War, showing the wide and low front along the Julianalaan. (taken, in adapted form, from the repository *"Beeldbank TU-Delft"* (*Delft University of Technology. Creative Commons BY*). **b** The main entrance (slightly adapted; *original photograph by Jose Krulic; Science Centre, Delft University of Technology*).

highly ramified. Each department had more or less its own branch of the building and also thereby could well develop and maintain its own atmosphere, without much interference with the other departments. But the laterally extended constitution of the building involved that it took some time before one could find its way effortlessly, a "problem" for visitors.

As the study progressed, the contact with real research became gradually more intense, as "pre-cooked", practical assignments became less. Instead small research projects had to be executed, which satisfied me much more. In the later study years also a considerable amount of free choice of offered lecture courses ("capita selecta") was possible. I well remember a course on statistical thermodynamics that was very much of my liking.

The Virtue of an Old Memory; The Hartman-Perdok Theory

One further example of these "capita selecta" was the lecture course given by Piet Bennema (with nickname "lange Piet", as he was rather tall (=lang)). His topic was "crystal growth". Bennema had completed a thesis in 1965 presenting (one of ?) the first application and test of the Burton-Cabrera-Frank theory for crystal growth, as controlled by "steps/ledges" in otherwise smooth surfaces of a crystal, as due to dislocations with screw character that intersect the surface: the "spiral growth mechanism". He joined the TUD in 1969 as a "lector" (comparable to an associate professor). I attended his lecture course on crystal growth in probably 1970/1971. At that stage Bennema was clearly not an experienced lecturer for students. His lectures were chaotic; he jumped from topic to topic (at least it made this impression to me, having no background in the field concerned) and regularly he did not even conclude his sentences… But that was largely compensated by his enormous enthusiasm.

It was the time that computer simulations emerged as an important tool in science. Bennema was involved in computer simulation of crystal growth together with George Gilmer, who was on sabbatical leave from the USA. They evidently performed groundbreaking work. A friend of mine and fellow student, Durk van Dijk, did his master project on such a project in Bennema's group and I followed this work, but was not involved. It was a most rewarding experience for me.[5]

Now I arrive at the point why all this is recounted here. In his lectures Bennema also discussed the Hartman-Perdok theory for predicting the morphology (the external shape) of crystals. He had spent time with both

---

[5] Clearly Bennema was too good to stay on a lector position. He accepted a Chair at the University of Nijmegen in 1976; a loss for the TUD.

Hartman and Perdok at the University of Groningen, so he had first-hand knowledge. This theory of 1955 aims at finding the relation between the crystal structure and the growth morphology of freely growing crystals.[6]

The theory could apparently be applied successfully to crystals growing from and in liquids. This did not have any relevance for me and I forgot about it for a very long time.

At the end of my career, at about 2015 in Stuttgart (so more than 40 years after the time of the narrative presented above), I and some of my co-workers were studying the "whiskering" phenomenon, involving the intriguing, seemingly spontaneous development of "hairs" on top of a (metallic) surface (see Fig. 5.3 and "*Whiskering*" at the end of Chap. 13). In our case it concerned the outward growth of "hairs" of tin (Sn) from and on top of a Sn film. They have diameters typically in the range 0.5 to 10 μm (1 μm $= 10^{-6}$ (one millionth of a) meter) and lengths up to several hundreds μm (100 μm $= \frac{1}{10}$ of a millimeter). These filamentary needles are usually single crystalline. We were experimentally able to determine the crystallographic growth direction (=whisker axis) of the needles. The question emerged if we could understand why specific crystallographic directions occurred as whisker axes and others not. I don't know how it happened, but suddenly, like a lightning bolt from the blue, I remembered that I had long ago learned something about the Hartman-Perdok theory and I now believed it to be of relevance for our problem. For my co-workers this theory was totally unknown. But by studying it we were able to develop a modified application of this theory to our problem of whisker growth and obtained a very good agreement between our theoretical predictions and the experimental results.[7]

---

[6] The starting point of the Hartman-Perdok theory is the identification of chains of strong bonds running uninterruptedly through a crystal. These chains are called Periodic Bond Chains (PBCs); they are subdivided in building units (e.g. atoms), where each building unit comprises the minimal amount of atoms/bonds to form a repetition unit of the PBC. The starting and end point of such a building unit define the so-called PBC vector. The treatment then culminates in the conclusion that so-called F (flat) faces, which are faces of the crystal at least containing two co-planar PBC vectors, are the growth morphology determining faces, as these faces release the least energy upon attachment of a building unit (e.g. an atom) and therefore are the faces experiencing slowest growth on top of them.

[7] This modification can be presented as follows. Tin (Sn) whiskers grow from the base from surface grains with grain boundaries, in the tin film, inclined with respect to the surface (see Fig. 13.12). If these grain boundaries are immobile, the morphology of the periphery of the growing whisker rod is determined by the neighbouring grains of the whiskering surface grain. As a consequence, the desired for F faces (see footnote 6) cannot develop at the outside of the growing whisker (this is the decisive difference with a crystal growing freely within a liquid). If such F faces would develop, they would share one common direction: the whisker axis (in crystallographic terminology this common direction of the F faces is called the "zone axis" of these F faces). This whisker axis (zone axis of the F faces) then likely is a PBC vector shared by all these F faces (of which each contains a second PBC vector; see footnote 6). Then, although the desired for F faces cannot develop in view of the immobility of the grain boundaries of the surface grain from which the whisker originates, it

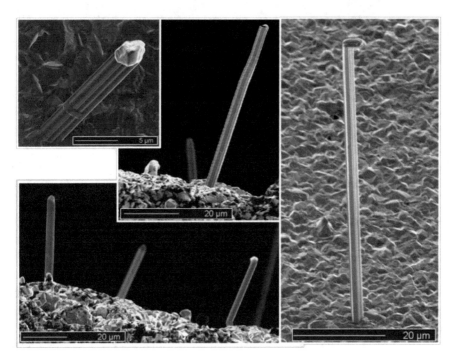

Fig. 5.3 Examples of tin (Sn) whiskers grown outwardly from the surfaces of tin films electro-deposited onto copper (Cu) substrates. Scanning electron microscopy (SEM) images of specimens prepared with a focused ion beam (FIB) workstation (1 μm = 10^{-6} m; taken from *J. Stein, Ph.D. Dissertation, University of Stuttgart, 2014*).

This late experience of utilization of a recollection of knowledge, until then residing "sleeping" somewhere in my brain, has impressed me greatly. Science progresses by building up on knowledge acquired in the past, and by, perhaps especially, making the "bridge" to the research currently performed. This is not a trivial process.

Whereas in the first two years I had travelled by train more or less daily from my parents' house in Haarlem to Delft in the morning and vice versa in the late afternoon/evening, I moved to a small one-room student apartment in Delft at the Roland Holstlaan at the end of my second year. It was difficult in Delft to find an appropriate student apartment, but, now having had two

is likely that the whisker axis yet is parallel to a PBC vector, as a result of the strived for but not realizable morphology. On this basis agreement of thus predicted and observed whisker-axis directions was obtained (*J. Stein, U. Welzel, A. Leineweber, W. Huegel and E.J. Mittemeijer, Acta Materialia, 86 (2015), 102-109*).

years of commuting and good results in the study, I had a high priority for being allotted such a reasonably attractive apartment. So, with 19 years old, for the first time I was more or less completely independent, which suited me well.

Actually I was not so "independent" as the above may suggest. Financially I was fully dependent on my parents. This was a significant offer my parents, having modest financial means, were prepared to make. Also therefore I felt obliged to finish my study as soon as possible and, while studying, with at least sufficiently high marks so that I would be freed of paying tuition fee (which in any case was modest in The Netherlands at the time). As a consequence my possibilities to party as a student were limited; more than a pub visit with friends, once a week or so, was too expensive.

The Roland Holstlaan is a street in an, at the time, relatively new quarter of Delft with the name "Voorhof". The relatively small street was aligned by house blocks of eight stories. No garages were provided for, so an overwhelming amount of cars was parked on the street, which didn't add to the appeal of the street. The flat with the student apartments in this street was exceptionally high: it had about 18 or so stories. The huge building was supported by a construction of steel, which implied such mechanical flexibility that during a storm the highest part of the building moved measurably, laterally, with the wind over various centimeters, as for example revealed by lamps hanging from the ceilings at the upper floors. Nearby a supermarket and a laundry service were situated, so a student could "survive" here…

One of my neighbours was a mathematics student. He had a pleasant personality. We developed a friendship. During a couple of years we had many long and pleasant discussions, also about science, while drinking a beer or so. He was apparently always glad to see me. His apartment was loaded with newspapers and other reading stuff, which emphasized the chaos in his room; when I visited him he had to move things away so that I could sit. He did not study intensively, if at all. There was some remark by him of having a problem with somebody of the mathematics faculty. I tried to stimulate and push him a little to solve "the problem" or finding a way circumventing it. Instead he was following the stock market closely and was betting on stock prices. Later I had to conclude he was really dangerously addicted to this. He was not successful with this activity and also not in his study. He seemed to get nowhere. Quite a number of years later, having moved away from the students' flat, I had already obtained my doctor's degree, I saw him one last time, passing me in opposite direction on the street. He also noticed me, but avoided any contact. I have not forgotten this former friend and wonder what has become of him.

A decisive experience was my confrontation with solid state chemistry/physics as presented in the lecture course given by, then already emeritus, Professor Burgers (see "*My contact with Burgers*" in Chap. 10) and an (X-ray) diffraction course that was given in his former group, now led by his successor Professor Boudewijn Okkerse, in the Laboratory of Metallurgy at the Rotterdamseweg 137, close to the "Gele Scheikunde". The diffraction phenomenon, which can reveal details of the atomic arrangement in (crystalline) material, and its power to identify and quantify mistakes in the atomic arrangement of (solid) material, captivated me. Especially the recognition that solid material, as viewed by the human eye, may seem to be in rest but in reality undergoes transformations by rearrangements of the atoms within, possibly, periods of time comprising seconds or years, fascinated me, also because such processes run in materials of great importance to mankind (e.g. steels). This led to my decision to perform my Master's project ("afstudeer-project" = graduation project) in the group of Okkerse in the Laboratory of Metallurgy. I didn't realize at the time, but this decision meant that I would become and eventually remain a materials scientist for the rest of my career.

My move to the Laboratory of Metallurgy for my Master's project involved as well that I got a few more friends, who also did their Master's project in Okkerse's group, but they had followed the academic study course "Metallurgy" from the beginning.

One day I and one of these new friends wondered why there had to be lavatories which were allowed to be used exclusively by the Professors, as indicated by a small shield on the door of these lavatories. So, one night both of us removed those shields. The following days nothing happened. Somewhat later a new shield was mounted on the doors of these lavatories indicating that these were now "general" lavatories accessible to everybody in the laboratory. Apparently, the board of the laboratory had accepted that such "class" distinctions for the use of lavatories did no longer hold up with the time in which we lived…. I have remarked already earlier that The Netherlands is a country of strongly egalitarian nature (see Chap. 3).

The specific topic offered to me in Okkerse's group, and that I accepted, dealt with the analysis of concentration profiles by (X-ray) diffraction analysis.

Diffraction Analysis as a Tool for Determining the Constitution of Matter; Regularities and Irregularities of the Atomic Arrangement

As one may recall from physics as taught on secondary schools, when a beam of monochromatic light (associated with a wave phenomenon) hits a grating of *equally spaced* slits, the interplay, i.e. interference, of the secondary waves originating from the separate slits can give rise to a pattern of bright and dark lines on a screen at relatively large distance from the grating. This pattern is called a diffraction (interference) pattern. The point here is: the wavelength of the incident light must be of the same order of magnitude as the distance of the slits in the grating in order that a full diffraction pattern can arise. The atoms in solid, crystalline materials can act as the slits in the grating referred to above, i.e. upon being hit by an incident beam of "light" they produce secondary waves. As the atoms in crystals are positioned at constant distances with respect to each other, the crystal has an internal regularity and thus can be conceived as a three-dimensional grating. Consequently a three-dimensional diffraction pattern can develop in principle.

However, with reference to the above remark, regarding the development of a full diffraction pattern from the one-dimensional grating of slits, the wavelength of the incident light must be of the same order as the distance between the atoms in order that a full diffraction pattern of a crystalline material can be observed. These interatomic distances are of the order of an Ångstrom (= 1 Å = 0.1 nm (=$\frac{1}{10}$ nanometer) = 10^{-10} m). The needed "light", that has a wavelength of this order of magnitude, is that of X-rays. X-rays were discovered in 1895 by Wilhelm Röntgen. Max von Laue was the first to recognize that the passage of waves of wavelength of the order of the distance between the atoms in the crystal must give rise to a diffraction pattern and that X-rays provide the appropriate wavelength and thus can be utilized to this end. Together with his co-workers, Walter Friedrich and Paul Knipping, he proved experimentally the correctness of this idea (1912). Von Laue at the same time also presented the quantitative diffraction theory (nowadays called the "kinematical" diffraction theory) to describe the effect. It was William Lawrence Bragg who shortly thereafter, in the same year (1912), presented his now famous, easy to derive law, in fact a geometric construction, that explains the positions of the intensity maxima in the diffraction pattern (1912),[8] which positions fully agree with those obtained from the description of the entire intensity profile for each diffraction maximum as given by von Laue.

Diffraction analyses of crystals were and are foremost used to determine the crystal structure of substances and also, reversely, for material identification using the diffraction pattern as a "fingerprint" of the material investigated.

---

[8] Bragg showed that the positions of the intensity maxima can be conceived as the result of "*reflections*" of the incident "light–waves" (X-rays) *from the atomic planes* within the crystal. Bragg's law (as it is called now; see the remark about "physical law" in Chap. 4) reads: $n\lambda = 2d \sin\theta$, where λ denotes the wavelength of the incident radiation (X-rays) that "hits" the atomic planes considered at an angle θ, d represents the spacing of the set of atomic planes considered and n is a positive integer (called "order of reflection"). Provided Bragg's law is satisfied, an intensity maximum occurs in the "reflected" direction, i.e. also at an angle θ with the atomic planes considered.

Already briefly after the discovery of the diffraction of X-rays by crystalline materials it was realized that crystals are far from perfect considering their atomic arrangements: crystals can be small (i.e. not infinitely large[9]) and can exhibit defects in the atomic arrangement. These effects lead to broadening of the diffraction maxima/lines. Scherrer (1918) and Dehlinger and Kochendörfer (1927, 1939)[10] published the first seminal works concerning the analysis of the (small) crystal size and the mistakes in the crystals from the corresponding diffraction maximum/line broadening.

A material in rest and in the absence of external loading can yet be subjected to, even huge, internal stresses. Such stresses can result after some (e.g. heat or mechanical) treatment of the material/component considered and they are then called *residual* stresses. The endurance life of such materials/components can be strongly influenced by the presence of both residual (internally applied) and externally applied stresses. Obviously, the presence of these stresses cause changes of the (average) distances between the atoms and consequently, following Bragg's law (see footnote 8), the diffraction maxima will shift (i.e. occur, for constant λ, at other values of θ; cf. footnote 8). Glocker[10] published in 1927 a textbook in which the basis of the (X-ray) diffraction method to determine the stress, from such diffraction maxima/line shifts, is described.

Hence, small size of the crystals constituting a material, mistakes in the atomic arrangement and (residual) stress are defects which can be decisive for the performance of a material and can be analyzed, more or less exclusively, by the broadening and shifts of the diffraction maxima. This explains the importance of these methods for analyzing of what is called "the microstructure of a material".

The application of (X-ray) diffraction played a very large role in my research during my whole career (see also Chaps. 10 and 13). In my graduation and following Ph.D. project the focus was on determining concentration profiles within materials by diffraction methods. Concentration variations are normally coupled with variations in the distances between the atoms and thus in general lead to broadening and shifts of diffraction maxima....

---

[9] At first sight it seems odd to conceive the finite size of a crystal as a defect. However, considering properties of ideal crystals, the influence of the surface of crystals is not taken into account, i.e. infinitely large crystals are considered then. And, indeed, as defects in the atomic arrangement can do, smallness of crystal size (in practical terms, say, smaller than 1000 Å (=100 nm = 0.1 μm = 10^{-7} m), does give rise to measurable diffraction-maximum/line broadening.

[10] Dehlinger and Kochendörfer and Glocker did their work at the University of Stuttgart, which would also be my affiliation (in conjunction with my position at the Max Planck Institute for Metals Research) during the second half of my career.

The interest in the X-ray diffraction analysis of concentration variations arose on the basis of a theoretical paper by a Professor in the USA. The topic was thought to be something for a graduation project. Now the paper was full of, for me at the time, complex calculations, which were not fully understood at once. Evidently I needed distinct background on diffraction theory to come forward here. I started studying the book "X-ray Diffraction" by B. E. Warren, which I did in my small apartment. As I didn't show up for at least a week at the laboratory, people there thought that something had happened to me and were more or less relieved by my reappearance. Apparently my solitary line of approach to my first genuine research project was considered unusual. Yet, it paid well off for me. I had learned a lot (the book by Warren is very good) and began to understand, step by step, what happened in the paper studied. In the course of time I was able to demonstrate that the entire paper, which focused on the description of the shape of diffraction profiles/maxima in the presence of a concentration profile in a crystal, could be much simplified by definition of a single factor which characterizes the shape of the diffraction profile in all cases. Thereby those cumbersome calculations in the studied paper had become unnecessary. Some time after my graduation this result was published as a short communication; this, being one of my very first steps in the world of professional science, made me a little proud and in any case some self-confidence resulted, which is the reason to recount this story here.

The project not at all was mainly a theoretical one. Diffusion couples had to be prepared (thin layers of copper (Cu) electrodeposited onto a single crystal of nickel (Ni)), diffusion anneals to be performed, the X-ray diffraction experiments had to be executed and subsequently to be analyzed.

Both the above theoretical activity and the analysis of the experimentally determined X-ray diffraction profiles required involvement of more or less complicated calculations which could not be performed analytically and not by hand. Consequently I spent a lot of time during my Master's project (and following Ph.D. project; see what follows) at the computing centre of the TUD, implying that I regularly walked up and down the Prins Bernhardlaan, connecting the Laboratory of Metallurgy and the computing centre, carrying large stacks of punched cards. As compared with nowadays practice, this can be interpreted as the "Stone Age" of computational science. This impression can be even enhanced by noting that the central computer of the TUD was an IBM 360/65....; present-day laptops have far greater computational power.... At the time this IBM machine was state of the art and indispensable for me. I still remember the enormous satisfaction I felt once I was able for the first time to deconvolute[11] successfully our measured diffraction spectra, which had required a long time of programming and debugging.

The Master's project was successfully concluded. I had indeed spent not more than the 5 years to complete the chemistry study, as foreseen in the study program. Of the about 165 students who had started in 1967 only 2 managed accordingly.

In the end phase of the Master's project, the possibility for continuing the project, but then as a Ph.D. project, was extensively discussed. Okkerse found this a good idea and provided for me the funded Ph.D. student position.

As a Ph.D. student I got a modest income and from now on I was financially independent. The following years were rather good ones. I concentrated on my research, having no great other "problems" or obligations that could distract me. One Professor once told me that I should consider these Ph.D. research years as singular, as later in life such uninhibited devotion to research would no longer be possible, as additional responsibilities would impose constraints. I would find out that this was all too true.

As a pupil at the Lorentzlyceum, where I followed the HBS-b, I found that in the final year, preparing for the concluding examination, I had to work hard and considered the last months to be very stressing. I naively thought that expenditure of such amounts of energy and stress would be a one-time-in-life occurrence. However, this experience would be part of my life in a repeated, crescendo manner: approaching the end of my Ph.D. project, I was working more and more and got accustomed to working at night as well. Later, as a Professor leading a large department, the work load was even larger. Although I liked most of my work (the science and teaching-mentoring part) very much, which made all input from my side bearable, I sometimes had to slow down, as I reached the end of my capacity and had to protect my private life. These remarks are made deliberately: not to put off somebody of a career in science, but to make clear that for success in science great devotion and love for science and colossal energy are required; it is not a "job" that can be done "from 9 a.m. till 5 p.m.".

Already during my Master's project, I got to know Kees Brakman, who also was a Master's student at that time. He was a metallurgist. For his Master's project as well as his following Ph.D. project he was prepared to go "to the limit" (for his Master's project he had written the bulkiest Master's thesis in the history of the faculty). Actually that was a characteristic of him fitting

---

[11] Deconvolution is a mathematical operation, in this case carried out in Fourier space, that allows to "remove" the broadening of instrumental effects (e.g. as caused by the width of slits through which the X-ray beam passed) from the measured diffraction profiles, so that the resulting profiles exhibited only the broadening due to the physical effects to be studied (e.g. the broadening caused by mistakes in the crystal structure; see the above intermezzo *Diffraction Analysis......*).

to almost all aspects of his life, which therefore was not always easy for him. We became friends. Often we went in the evening together to the Chinese restaurant at the start of the Rotterdamseweg, i.e. on walking distance from the Laboratory, enjoyed our meal and returned later to the Laboratory to continue our work. One time at the Chinese restaurant we suddenly noticed a cockroach crawling on the wall adjacent to our table. These insects are inno-cent for men, but repellent to see. Their presence indicates that the kitchen of the restaurant is not hygienic. This did not stop us visiting the restaurant again (there was no alternative nearby); we were young and a bit reckless…

Kees had a strong appeal for complicated mathematical equations. His topic was the analysis of texture (and stress) in solid materials. With the notion "texture" in materials science the preferred orientation of the crys-tals in a polycrystalline specimen is meant. This description, by a function called the "orientation distribution function ("*odf*"), requires application of cumbersome mathematical functions. Such a complex function, with multiple integral and summations signs stayed, for at least a couple of years, on the blackboard in his office; it was categorically not allowed to be wiped away by the cleaning lady or others. To see it, what could not be avoided upon entering his office, of course impressed his visitors, which I presume was a desired side effect, but Kees really had distinct mathematical skill. Years later Kees published a short note with the rather nonattractive, opaque title (the reader does not need to understand the mathematical symbols in this title): "*Evaluation of the integral* $\int_{-1}^{1} P_l^{m,n}(x) P_l^{m-2,n}(x)dx$ ",[12] where $P_l^{m,n}$ are generalizations of "associated Legendre functions", all related to the *odf* mentioned above. In this paper, *by conjecture*, Kees arrived at the solution for a certain nontrivial summation (Eq. (11) in the paper), which in fact was what the whole paper was about. I found and still find this a remarkable sign of genuine insight. The evidence that Kees' guess was correct was given in a later paper by two mathematicians[13] and required 4 pages of print….

As noted, my Ph.D. project was an extension of the Master's project, but in the course of time the scope was much broadened. Not only single crystalline diffusion couples but also polycrystalline powder specimens were subjected to diffusion annealing and X-ray diffraction analysis. Research on methods for the determination of the concentration dependence of diffusion coefficients from determined concentration profiles led to separate papers. A significant step was the decision to investigate the effect of interdiffusion in the trans-mission electron microscope employing as specimens thin films, in particular

[12] C. M. Brakman, *Textures and Microstructures*, 7 (1987), 207–210.
[13] H. Bavinck and R. F. Swarttouw, *Textures and Microstructures*, 10 (1988), 37–40.

copper/nickel bilayers. This was a new area, where we developed a method to deduce the diffusion coefficient from the broadening of the spacing of so-called moiré patterns upon annealing.[14]

The investigations were partly done together with Master's students, a few of them being former fellow students of mine. Without them it would have been impossible to bring research, in so many subprojects as indicated above, to success in relatively short time.

The very first time that results of our work were presented in public was on the occasion of the 10th IUCr (= International Union of Crystallography) Congress in August 1975 in Amsterdam. We had three posters at this meeting. It was the first time I saw and met scientists famous in our field:

Thus we had a lengthy discussion with J. (Jerry) B. Cohen, the editor of the Journal of Applied Crystallography. Some years later discussion with him would be continued (see Chap. 10).

I attended presentations by William (Bill) Parrish, who spent a whole career on perfecting instrumentation for the diffraction analysis of crystals and who was the famous developer of the vertical Philips diffractometer. He was a person with a loud voice and an impressive, intimidating personality. It was not easy to confront him critically in a (public) discussion.[15]

---

[14] Moiré patterns, as meant here, are caused by interference of two (diffracted) waves, one from a thin crystal of nickel (Ni) and one from a thin crystal of copper (Cu) on top. The crystal structures of copper and nickel are the same, and here have the same orientation in space, but have slightly different lattice parameters, so that at their interface a "lattice mismatch" occurs (cf. two line gratings on top of each other, oriented identically (i.e. the lines of the gratings are parallel) but of slightly different spacing). The bicrystal is such thin (say, thinner than 1000 Å = 100 nm = 0.1 μm = 10^{-7} m) that it is transparent for an incident electron beam (in the electron microscope), which can be conceived as a wave phenomenon (cf. the dualistic nature of light). This incident electron beam is diffracted in both crystals during its passage through the bicrystal. The orientations in space of a pair of similarly diffracted electron beams (one from copper and one from nickel) are close to each other. Image formation with such a couple of diffracted beams (by application of an appropriate aperture in the microscope) causes an image revealing a contrast composed of a set of dark and light lines *with a spacing determined by the lattice mismatch (lattice-parameter difference)* of copper and nickel (this difference for pure copper and pure nickel is about 2.6%). Upon annealing atoms of copper diffuse into nickel and vice versa. The lattice mismatch thereby decreases and this causes an increase of the spacing of the moiré pattern. On this basis diffusion-coefficient values were obtained at exceptionally low temperatures, which is an important advantage of this in situ method (*T. van Dijk and E.J. Mittemeijer, Thin Solid Films, 41 (1977), 173–178*).

[15] Four years later, one year after I had completed my Ph.D., Staf de Keijser (see Chap. 10 under "*Diffraction line profile analysis; the Voigt function and the "Dream Team"*") and I attended the "Symposium on Accuracy in Powder Diffraction", organized by the National Bureau of Standards (NBS; now renamed as National Institute of Standards and Technology, NIST), in Gaithersburg, Maryland, USA (June 11–15, 1979). Both of us had a big lecture to present. Bill Parrish was also there. Our work was not really criticizing Parrish, but also not supporting his approach. Mentally I was prepared for a question, also a possible critical one, by Parrish in the discussion after my lecture. I had not even completely finished speaking or the massive body of Parrish already rose from his seat. Before the chairman could have given Parrish the opportunity to pose his question or comment, I loudly said,

And then there was Paul Peter Ewald, the contemporary in Munich of von Laue and close to the discovery of the diffraction of X-rays himself. Ewald, a world famous, great scientist, was the one who introduced the concept of the Ewald sphere in the reciprocal lattice and pronouncedly contributed to the theory of dynamical diffraction. At the time of the Amsterdam meeting Ewald was already 87 years old. This did not withheld him from moving with great dynamics from lecture to lecture dressed in shorts, revealing his really old legs. From 1928 till 1938 Ewald had been Professor in Stuttgart, my future affiliation some 23 years after the conference. Ewald left, more or less forced, Stuttgart in 1938, as so many scientists in disagreement with and who criticized the political, NSDAP/Hitler system. He first went to Cambridge (United Kingdom), then to Belfast (Northern Ireland) and finally in 1949 to Brooklyn in the USA (also see Chap. 13).

Approaching the end of my four years term I requested Okkerse to extend my promotion time with one additional year. I had the correct feeling that this would allow to complete a number of papers and thus make the thesis more comprehensive. Okkerse agreed and thus at 22–2–1978 the thesis was defended successfully in public before a committee that also included Professors Burgers and Meijering; I will speak of both in a later chapter.

From Science to Management; The Career of Professor Okkerse

My promotor, Professor Boudewijn Okkerse, left the university about one year before the end of my term as Ph.D. student, to accept a high ranking position at the Ministry of Education and Science in The Netherlands: he became the (first deputy-, then) Director-General at the Directorate General for Higher Education and Scientific Research, a position directly under the Minister. This did not hinder him to remain my promotor, because a Professor, after having left office, kept the right to act as promotor for three years beyond. At the Laboratory we were not really shocked by this step; some such move was almost expected (see what follows).

Okkerse had been a successful Ph.D. student (see Chap. 10) and also his years at the Philips Research Laboratories (=Philips Natuurkundig Laboratorium; short notation: NatLab) had been a success: he performed beautiful experiments, later cited in textbooks about X-ray diffraction, on the anomalous transmission of X-rays, using extremely perfect (germanium) crystals grown by himself. The associated theoretical work was largely in the hands

---

actually it gushed spontaneously out of me: "*Oh, oh, here it comes!*". The audience laughed, recognizing the confrontation of an established, famous, much older scientist with a relatively unknown young man. Parrish was at once unarmed and the stress I felt relaxed. His following question was easy to respond to.

of his colleague Paul Penning, co-originator of the Penning-Polder equations in the dynamical diffraction theory, who became, before Okkerse, a Professor at the Laboratory of Metallurgy. After some years at the NatLab, Okkerse (was) moved (cf. the corresponding remark about the personnel policy at the NatLab in the main text below this intermezzo) to the Philips Electronic Components and Materials Division (Elcoma). He became a manager (director of the development laboratory of Elcoma) and no longer performed research. After the retirement of Professor Burgers in 1968 he was approached to become the next occupant of the Burgers Chair. And thus in 1970, as a chemistry student, I attended his lectures on physical chemistry of the solid state, basically presenting some crystallography and especially the character and role of crystal-structure defects.

I don't know what the original ideas of Okkerse were regarding his role as Professor in the Laboratory of Metallurgy. He was very well organized and disciplined, held regularly meetings with every co-worker and Ph.D. student, who had to write reports about the progress made since the last meeting. Okkerse lacked a distinct vision on how to develop the science performed by his group. Admittedly, it may have felt difficult for him to drastically change the course of the activities of especially his elder, senior co-workers, who all had own ideas of what to do and more or less continued what they already did under Burgers. He may not even have desired to intervene pronouncedly, as the quality of the research performed was generally high.

In fact Okkerse remained the "manager" he was in his last job within the Philips company. The winged words often expressed by him were: "*At Philips we then did......*". It is likely in general very difficult for somebody coming from industry and having pronounced manager experience, having successfully shaped his/her professional environment and thus steered many people, to return to, in view of the challenges for a manager in an industrial environment, a relatively small group of people working on apparently small, seemingly insignificant scientific questions, as worked on oneself in earlier years as a Ph.D. student and young co-worker. Such a person may have lost the affinity for science as a vocation and mission. I believe this is what happened to Okkerse. In fact the same held for many a Professor at the TUD at especially the core engineering faculties, where Professors were often appointed after they had been manager in industry for years: often their subsequent scientific performance at the university was meagre.[16]

---

[16] Very recently, while drafting this chapter, I came across a published interview with a Professor, on the occasion of shedding a full professorship at a university of technology for a new job as full manager of an institution of higher education (below the level of a university). Upon the question why this move was made, the following response was given (translated into English): "I was just one

Okkerse was a decent man. He accepted his modest role in the scientific output of his group. He never claimed or wanted to be a co-author of any paper from his co-workers and Ph.D. students. There was one exception, where one co-worker, working on a problem inherited from Burgers' time, had a personality that was an obstacle in finishing his project. Okkerse helped, brought, with his assistance and especially push, the project to an end and indeed then fully justifiably appeared as co-author on the single paper published.

His unpretentiousness was also shown by his choice of office: he did not take the spacious office once occupied by Burgers, but a small cabinet of half the size was used by him. But this had a serious disadvantage. If a larger, e.g. staff meeting was held there, scarcely all people could find a place in this narrow office and smoking there was impossible. All other Professors at the Laboratory of Metallurgy of course had a large office as Burgers had. Years later, upon my return to the group as Professor, I took the former Burgers' office and, by the way, although by then being non-smoker myself, allowed smoking in my office....

As a Ph.D. student I was too far away from the Board of the Laboratory of Metallurgy to know what precisely went on there. In my memory there were rumours that some sort of reorganization of the Laboratory was proposed by Okkerse, which was not well received by his colleagues. Okkerse may have felt obstructed in his management ambitions. This may have come on top of his realization that he did not function as a Professor/scientist in the way as may have been his intention upon accepting to become the successor of Burgers. Feeling unable on both fronts may have led to his decision to become the (deputy- then) Director-General at the Ministry of Education and Science. In a way, and among many other responsibilities, he thereby became the person in charge of all universities in The Netherlands...; *that* may have pleased him.

Characteristically for Okkerse, he immediately dropped his title "Professor" in his new job and privately. On the eve of the defense of the Ph.D. thesis, Okkerse proposed I should call him "Boudewijn". I couldn't and for a long time afterwards I kept addressing him as "Professor". Only much later,

---

year no longer Dean of the faculty and felt that I actually missed the managing. Managing I like more than I thought previously......I am going to manage really fulltime. I had to think thoroughly about that. But I really look forward to it." (*De Ingenieur, September 2020*; "De Ingenieur" is the monthly journal of the "Royal Netherlands Society of Engineers"). This is just a more than evident illustration of what I have implied above: Having experienced the taste of exerting a powerful, controlling, management function ("ruling" over many people), the zeal for science and scientific research (with a comparatively small group of people....) can easily get lost.

being older, more mature and a Professor myself, I felt confident enough using his Christian name.

Okkerse stayed 15 years at the Ministry. At the age of 62, a reorganization at the Ministry occurred, apparently affecting his field of activities. He resigned.

In later years, at the times that we had contact in person or via mail (see Chap. 10), I had thought asking him possibly about his motives for his decisions to leave the TUD and later the Ministry, to confirm what I presumed (and hinted at above). In the end I decided not to do that.

The thesis was composed of published and at that stage still unpublished, submitted papers. This was new for the TUD. A change in the rules for promotions at the TUD had happened briefly before, thereby allowing to compose a Ph.D. thesis from published and as yet unpublished papers. Until then the custom was that, before one or more papers were written on the basis of performed Ph.D. research, a small or larger booklet was written (often not in English, but in Dutch) with a lot of text of often limited interest (long introductions and long discussion of literature and possibly also minor results not mature enough for publication in a journal). After the Ph.D. examination then a second writing task had to be carried out; one or a few papers were extracted out of the thesis, i.e. condensed rewriting (often in a different language (English)); only a fraction of the thesis contents usually concerned publishable material. This is a very inefficient, laborious approach to disseminate hopefully important, scientific results, not only for the Ph.D. candidate, but also for the supervisor and promotor. Moreover, often the Ph.D. candidate usually leaves the university after the promotion and depending on his/her job thereafter may not have the time or interest to focus on such "paper writing".

I have always strongly favoured the "cumulative" model of thesis writing indicated above: all theses by my later Ph.D. students, in Delft and Stuttgart, were composed of their papers (as published or before submittance) as the chapters of their theses and, of course, written in English, recognizing that English is the "lingua franca" of science (see also Chap. 13 under *Manners and Mores*).

Nearing the end of the Ph.D. time, it became time to think about the future. Two possibilities appeared before Eric's eyes: either to stay within the academic world, at least for some time to come, or to go for a job in industry. A strong tradition within Okkerse's group was to continue one's career at Philips at the Philips Research Laboratories (NatLab) in Eindhoven.

Not only Okkerse himself, but also the founder of the Chair, Professor Burgers, and quite a lot of his Ph.D. students (Okkerse had been a Ph.D. student of Burgers as well), had followed this route. As a matter of fact the Philips Research Laboratories in Eindhoven, founded in 1914, took a unique position: it was considered as the (former AT&T) "Bell Laboratories" (in the USA) of The Netherlands: one could perform there relatively free fundamental research, of course within the fields of interest to Philips, and publish results in the international literature. Its directors were, at least in the first seven or so decades of its existence, high caliber scientists, e.g. Prof. Hendrik B. G. Casimir[17] and Prof. Andries R. Miedema (see Chap. 13). Many a Professor at a Dutch university came, i.e. returned, from Philips to the university. Against this background many young physicist, chemist, materials scientist, usually after having acquired a Ph.D, longed for a position at the NatLab.[18] Eric's surroundings more or less expected that he would also wish a job there. Indeed he applied and was offered such a position.

Eric's appetite for scientific work had become only larger during his Ph.D. work. It was well known that at the Philips Research Laboratories, after being appointed as a young scientist, only a limited number of years was in general available for staying there. In fact the laboratories were considered within Philips as a breeding ground for future managers/leaders at other locations within Philips and this fate "hit" most young scientists joining the laboratories. This did not suit Eric's taste; he wanted to avoid such course of his future career. It happened at this moment of time that a vacancy for an assistant professorship (=wetenschappelijk medewerker) was available in the group of Professor Korevaar at the Laboratory of Metallurgy. This group had no high status within the Laboratory, in any case not comparable with Okkerse's group. To Eric the position was attractive yet: he saw the chance to develop his own group there, as Korevaar, in discussion with Eric, offered

---

[17] Casimir's name in science lives on in the "Onsager-Casimir relations", the "Casimir effect" and, in the field of mathematics, the "Casimir operator". Casimir was not only director of the Philips Research Laboratories but also a member of the Board of Philips, carrying the responsibility for the research of the company. As a retired person he wrote a recommendable memoir that illustrates a lot of the atmosphere and the academic and industrial science in The Netherlands (and the world) during his lifetime: *Haphazard Reality—Half a Century of Science*. Harper & Row, New York, 1983 (I read the Dutch version a couple of years later).

[18] In the eighties of the past century the character of the Philips Research Laboratories (NatLab) changed dramatically. Philips experienced a rather extreme decay and thus its research capacity, the NatLab, was cut and the character of research changed: research of fundamental nature practically disappeared (as it was believed that this no longer led to product innovations…) and the accent was put on product development in narrower sense. The NatLab and its reputation in fact dissolved. In these years Bell Laboratories followed a similar course of very severe cuts and a focus away from fundamental research to "immediately marketable areas", but recovered to a certain extent after about 2015 in the area of information and communication technology; since 2016 Bell Labs is part of the Nokia company.

him this freedom and the support of two technicians/engineers; Eric would have to apply for funds for Ph.D. positions. Also the resources of the Laboratory, including especially those of Okkerse's group, would be available to him; this also made continued cooperation possible in the area of diffraction-line broadening analysis (see Chap. 10). In short: Eric saw a chance to make a name by showing that he could develop and lead a (modest) group and producing high quality science. He felt that Korevaar would not interfere.

Hence, to the great surprise of Okkerse and others, I decided for the position in Korevaar's group. Especially I will not forget the astonishment and also joy of Korevaar, who knew about the Philips offer, at the moment I told him my decision. I never regretted this choice, as my expectations could be realized in the course of time. Korevaar had a very pleasant personality; he was modest (did not consider himself as an important scientist) and in fact supported me also later in all possible ways. I can only most gratefully remember my time with him.

Thus my student and Ph.D. time was concluded and a new adventure began in March 1978.

6

"Publish or Perish"; The "Growth" and Progress of Science

Abstract The increase in the number of publications from past to present, leading to the present deluge of published papers. Argumentation that the average quality of a published paper has deteriorated significantly. *Quantitative* analysis of published science does *not* inform about the progress of science, i.e. the growth rate of scientific understanding of nature. The positive and negative workings of the internet. Fraud and deceit in scientific papers becoming more prominent in recent years. Top journals, eager to realize the speediest publication of a result thought to be startling and/or epoch-making, are prone to publication of immature and even fraudulent work. The high status of a Ph.D. degree is associated with the increase of recent cases of plagiarism in Ph.D. theses. Experiences as Editor of an international scientific journal. The limited significance and the abuse of a single number, as the "impact factor (IF)", as a measure for the quality of a scientific journal and of a single number, as the "Hirsch (h) number", as a measure for the quality of a scientist.

In former days, scientists usually did not publish a lot. Even famous scientists at the end of their life/career often had an, in our nowadays view, modest list of publications (apart from exceptions: e.g. see the text about Charles Janet in Chap. 5 under "*Looking at The Periodic System; The Role of "Outsiders" in Science*"). My promotor, Professor Okkerse, once said that one publication per (man)year was already a very respectable achievement.

In that now distant past the published work was of more seminal character than is currently the case. The reputation of a scientist was not based on the number of his/her (unfortunately, mostly "his") published papers and it certainly did not necessarily hurt someone's standing if that number was rather meagre: A series of famous names in my field of science can be given for which this holds.

© The Author(s), under exclusive license to Springer Nature Switzerland AG 2022
E. J. Mittemeijer, *How Science Runs*,
https://doi.org/10.1007/978-3-030-90095-3_6

In those days it was also in practice unnecessary for a scientific career to have completed a doctor's (Ph.D.) thesis. And for those who did research for a Ph.D., this could become a "life's work". To have spent (even more than) seven years in order to complete a doctor's thesis was no exception at all (at least in The Netherlands).

Until about the time I started doing research as a (Ph.D.) student, the atmosphere in Okkerse's group was still largely characterized by the above description. In the USA (China was still absent in the scientific world) there was already a much more competitive attitude and the number of papers published became important as a quality characterizing magnitude. This drive crossed the ocean and gradually got the upper hand in Europe too.

Nowadays, the number of living scientists outdates the accumulated number of scientists who lived before us.[1] Already only this recognition suggests an exponential increase of the number of publications published. *Actually, this increase is distinctly larger than that expected on this basis* (see Fig. 6.1).[2]

Every currently living scientist feels the pressure to publish as much as possible in journals of possibly high reputation, as his/her status and career success depend on it to a high degree: "publish or perish". This has induced the publication of ill-considered or superfluous papers and also papers with purposefully manipulated or "created" data (see Chap. 12) and also papers with the incorporation of "co-authors" with impressive names who apparently did not bother to read and check the paper themselves and did not have any contribution of scientific importance. If the manuscript is not accepted by a

---

[1] The *UNESCO Science Report (2019)* recounts that the number of scientists worldwide has attained a value of about 0.1% of the world population, i.e. corresponding to about 8 million.

[2] The number of scientists doubles about every 18 years, using the number of Ph.D. degrees granted as indication (*W. Gastfriend* at https://futureoflife.org/2015/11/05/90-of-all-the-scientists-that-ever-lived-are-alive-today/). The growth rate of science, in terms of quantitative output, may be expressed as the number of publications (as journal papers) appearing per unit of time. However, the existing data bases, as Web of Science (WoS), cover only a part of the published scientific literature, which part moreover decreases with time (*P.O. Larsen and M. von Ins, Scientometrics, 84 (2010), 575–603*). Therefore the number of cited references per unit of time can be considered as an alternative. An increase in the number of cited references reflects the increase of citing and/or cited papers. On this basis, considering publications from 1980 till 2012, the growth rate in the last 50 years is found to double about every 9 years; see Fig. 6.1 (source: *L. Bornmann and R. Mutz, Journal of the Association for Information Science and Technology, 66 (2015), 2215–2222*).

It should be realized that in Fig. 6.1 the "growth" of science in the years before 1980 is mirrored by citing in papers from 1980 till 2012 (see above). Further, the flattening and then falling off observed in Fig. 6.1 for the years beyond 1980 is ascribed to increasingly fewer publications contained in the analysis (the part of all publications covered in the WoS decreases with time; see above). Moreover, those of the (especially newest/newer) publications which had not been cited yet are missed in the analysis. This last trend may even be more outspoken for publications in the field of materials science (see the comment about the "Cited Half-Life" under *"Editor of a materials science journal"* later in this chapter).

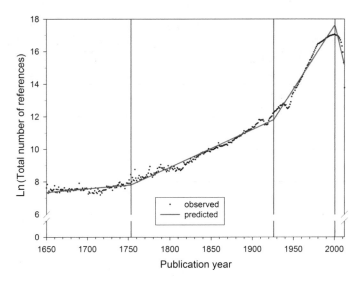

Fig. 6.1 The growth of the number of citations per year in the period 1650–2012. The data represent the references taken up in the publications from the period 1980 till 2012 as collected in a data base belonging to the Max Planck Society (based on the Web of Science (WoS) and maintained by the Max Planck Digital Library). Along the ordinate the number of references is represented by its natural logarithm (ln; e.g. ln 10000 = 9.21; ln 1000000 = 13.82). Focussing on the cited references in these publications illuminates the "growth" of science since early times. The strongest growth of 8% per year occurs in the phase that ends in this analysis with about the start of the 21st century, implying a doubling time of about 9 years (taken, in slightly adapted form, from *L. Bornmann and R. Mutz, Journal of the Association for Information Science and Technology, 66 (2015), 2215–2222*). The decrease in the (growth of the) number of cited references for years beyond about 2000 may simply reflect that (i) the number of publications not covered by the data base used increases with time and that (ii) a significant part of the (more) recent papers may have not been cited yet (e.g. papers in materials science are generally cited a first time only after some years have passed (see the remark about "Cited Half-Life" later in this chapter). The indication "predicted", corresponding to the full straight line segments in the figure, may generate the false impression of a *physical* model to explain the "growth" of science/number of references. These line segments represent only regression fits of arbitrary exponential functions (linear after taking the natural logarithm (see the ordinate in the figure)), which are simply used to describe the data as observed; i.e., this description is devoid of any physics.

high quality journal, often the same paper is sent to a journal of lower rank. In many cases the authors of mediocre or low quality papers may also directly address themselves to a journal of low quality. There are even many obscure, open access journals that accept any nonsense without any refereeing.

This all has led to the paper flooding we experience at present. Much of that work is of modest importance, as the tiniest and uninteresting results appear in (electronic) print. Also repetition of results published earlier has become "normal".[3] The results of a larger project are subdivided into small pieces, which subsequently are separately published, which is called "paper slicing", whereas one larger paper would have sufficed and would have been welcomed by someone interested in these results. Moreover, as indicated above, a pronounced trend to deceit and fraud in science developed (see footnote 5 and Sect. 12.2).

In view of the enormous increase in the number of papers published, one wonders how a scientist can keep informed about all progress made in his/her field of science. Given that the time necessary to read and study a paper has not changed over time, the described paper abundance of nowadays can only have as consequence that it can be a pronouncedly time consuming, even impossible task to detailedly inform oneself about really all what is published in one's field of science.

In the seventies and eighties of the past century, I went regularly, say once per week, in the library of the Laboratory reading the newest issues of the journals in my field (nowadays, of course, the number of journals in (also) my field has increased pronouncedly, in line with the above sketched development. Especially every new and hot subfield gets assigned its own, new journal(s), as this is a very profitable business for the publishing houses). Also, if the scientific problem I was working on needed that, I went to the library, retrieving, from the bound volumes of the journals in the library, the papers of my predecessors, by tracing back the references of relevance in their papers. This was a time and energy consuming task, but still possible to accomplish in these days. Nowadays an internet search (using the world wide web (www) service) may ideally provide all papers of any relevance for a certain scientific topic, but many papers of relevance are easily missed, as the search terms

---

[3] Not only publishing the same result by the same author(s) in a somewhat different context is meant here (exception: the publication of a full paper after a short note has already appeared; a usually acceptable form of this phenomenon). Especially, not referencing of earlier work by other scientists, who obtained the same or very similar results, is grave scientific misconduct. In that case either one was unaware of the earlier work (likely caused by inadequate literature research, if performed at all) or one deliberately did not cite the predecessors and then this behaviour is damnable vanity. By chance I have witnessed myself a few times that (i) substantial text of a paper written by me and co-authors was even directly copied (without reference to our paper), and that (ii) results were presented which by themselves were all right but in fact involved full reproduction of our earlier work that was not referenced at all. An especially mean trick is that in the case (ii) one yet does refer to the original work by other authors, but at an irrelevant place in the paper that does not make clear that the major findings of the present paper had already been published before by these other authors. Numerous times I have observed this. Also see the intermezzo "*Diffraction line profile analysis; the Voigt function and the "Dream Team*"" in Chap. 10.

Of course, I am not the first to notice this negative trend associated with the current paper flooding. An entertaining booklet dealing with this topic has been published recently (*G. Pacchioni, The Overproduction of Truth, Oxford University Press, 2018*).

used cause overlooking of these papers. However, and in any case, reading and studying them all is usually completely impossible.

The emergence of the internet at the end of the twentieth century has brought many advantages (e.g. literature provision and literature searches, but see above). It erased the slowness of the hard copy correspondence of former days, thus allowing direct (email) exchanges, (i) between colleagues at other places in the world, thereby enhancing international cooperations, and (ii) with editorial offices of scientific journals. However, at the same time a considerate way of manuscript preparation has suffered,[4] particularly at places lacking a high quality tradition. As was the experience in the past century, the labour intensive typing of a manuscript, where executing corrections (for example as response to referee comments) required complicated manipulations with the hard copy of the manuscript, and the time lost by sending manuscripts by surface or air mail to and fro editorial offices across the world, made it worthwhile to pay utmost attention to the science and associated (especially interpretative) text presented in the manuscript. Nowadays the practically direct communication via the internet has led to, in the above sense, more careless preparation of electronic manuscripts by authors of mediocre quality; some authors simply submit an early version of their work, wait for the referee comments, improve the paper accordingly and resubmit the paper (to another journal if the early version has been rejected by the firstly approached journal).

Also the number of Ph.D. projects increased with time: continuously more people embarked on a Ph.D. project (see begin of footnote 2 in this chapter). The time allotted for these projects became reduced considerably: the available (funded) time became four (The Netherlands) or three (Germany) years; in practice this became somewhat longer (five or four years, respectively).

In Germany and Austria there has always been a strong inclination to acquire a Ph.D., as in these countries the status lent by a Ph.D. degree means relatively (much too) much.[5] Thus I was amazed to observe that in

---

[4] As well as the "art" of letter writing. I regularly got emails from students, who hadn't written the note/request if they had spent a little more thinking time before addressing me. Moreover, receiving such emails starting with the salutation "Hello", from people who are not close to me, is bewildering. The handwritten or typed letters of the past had as distinct advantage that much more time, to contemplate what one wishes to bring forward, was involved before they were sent. This attitude would reduce a lot of the garbage sent via the internet.

[5] In recent years a series of Ph.D. theses, of especially but not only politicians in Germany and Austria, as members of parliament and ministers, were shown to be based to significant extent on plagiarism, causing deprivation of the Ph.D. degree. This grossly fraudulent behaviour attracted an enormous amount of public interest. This leaves undiscussed the lack of distinctly original, scientific quality of many Ph.D. theses, which, as it seems, particularly holds for dissertations in areas outside the natural ("hard") sciences (unsurprisingly, the plagiarism by the politicians meant in this footnote concerned Ph.D. theses *not* in the field of the natural ("hard") sciences).

the Faculty of Chemistry of the University of Stuttgart, of which I was a member after 1998, my colleagues considered only the doctor's degree as the "normal" conclusion of a chemistry study, which in The Netherlands, at least in the seventies, was not the case at all and as a result the number of Ph.D. degrees given was relatively small in that country. Instead, in The Netherlands, in any case before 1950, a Ph.D. degree was considered as an upbeat to a career in science... This is no longer the case.

In a way the universities have become "Ph.D. factories". A significant fraction of the Ph.D. students, as a consequence of the above sketched development, have no genuine talent for science; they only need the degree for facilitating (the entry to) a subsequent (e.g. management) career, in industry. This leads to the situation that a noteworthy fraction of the theses actually derive from the intellect and talent of the supervisors. This gives rise to the following type of remarks sometimes made towards these supervisors by their colleagues at the conclusion of such a Ph.D. project: *"You have successfully graduated another time"*. Such statement may be exaggerated, but, eventually, the phenomenon sketched in this paragraph will inevitably lead to degradation of the value of a Ph.D. degree in our society.[6]

It appears unrealistic to suppose that the type of exponential growth of science, in terms of the above parameters, can proceed endlessly. As a matter of fact the curve of increase in the number of Ph.D. degrees granted, for example, has flattened in recent years in Western Europe and the USA. But this has been (over) compensated by the rise of science in China.... An absolute, natural boundary to exponential growth of science seems to be the finite size of the world population, which is constrained by the resources on earth. But what happens if mankind starts to occupy other planets in the universe?

It may be clear from the above discussion that the measures used for assessing the growth of science have at best only quantitative significance for the specific parameters measured. They do *not* inform us about the progress of science, i.e. the growth rate of scientific understanding of nature. Therefore the word "Growth" in the title of this chapter has been put between quotation marks. Indeed, it appears more than likely, that the number of mediocre, or rather meaningless papers and of papers of low quality and of

---

[6] It is sometimes suggested to abandon the Ph.D. as an academic degree. The argument is that the quality of a scientist is judged on the basis of his/her published papers in the literature, not on his/her Ph.D. thesis which moreover often simply is a collection of such papers, published or (to be) submitted. However, this reasoning does not recognize that the completion and successful defense of a Ph.D. thesis, in the absence of negative developments as sketched in the main text above, represents an accomplishment indicating the ultimate completion of an academic education, comprising many years, *also for those, sooner or later thereafter, no longer pursuing a career in science*. In my view a societal sign of recognition, an academic degree, of this obtained level of competence is fully justified. Within this context, also see Chap. 13 under *"Manners and Mores"*.

even erroneous papers has pronouncedly increased in the last decades (see also footnote 3 in this chapter). This means that the average contribution to progress in understanding of nature, per paper published, has decreased since say 50 years, or so. Of course, this still does not imply that there is no progress in science and indeed very pronounced, even enormous steps forward have occurred especially in the last 200 years and this is a direct consequence of much more scientific work done by ever more scientists. However, in no way what is called "growth of science", as reported in scientometric/bibliometric papers, is proportional to genuine growth of science interpreted as increase in understanding of nature. This leads to a few remarks made on the basis of my experience as editor of an international scientific journal.

Editor of a Materials Science Journal

My last active years in duty and thereafter were also marked by my additional function as Editor-in-Chief of the *International Journal of Materials Research (IJMR)*, formerly called *Zeitschrift für Metallkunde*. This journal certainly is one of the oldest journals in our field of science: it exists for more than 110 years. The journal had initiated as an organ of the Deutsche Gesellschaft für Metallkunde (nowadays Deutsche Gesellschaft für Materialkunde). Its publications in the first decades of the past century were in German, also reflecting the world dominating position in science that Germany had at that time. Since 1959 papers appeared in English and the name change to *International Journal of Materials Research (IJMR)*, but only relatively late it might be said, in 2006, then is a natural consequence. The journal began as a metallurgy journal and, as consequence of the widening of the field of materials science, no longer is restricted to this field as also its current name indicates.

The journal has a large reputation. While in Delft I already had published in the journal, as did many groups from outside Germany, thereby indicating the international role of IJMR. Nowadays the impact of a journal is measured by bibliometric numbers rather than by its reputation as based on knowing the journal and being familiar with the type and quality of the papers published therein. An extensive scientometric/bibliometric analysis of IJMR in the past 100 years has been performed in 2009 by Werner Marx (*International Journal of Materials Research, 100 (2009), 11–23*), which gives rise to the following general and IJMR specific comments.

General comments:

– For a start it must be realized that the average citation rates in the field of materials science, as based on data given by the Science Citation Index (CSI), accessible under the Web of Science (WoS), is appreciably lower than in the fields of physics and chemistry. This likely indicates an intrinsic characteristic, but this observation can be enhanced by the tendency of the WoS to underrepresent the more technology/engineering oriented literature, where materials science papers are likely cited relatively often.

- Download counts from electronically available papers do not at all correlate with the citation counts for the same papers. Evidently, the first interest in a paper, as raised by a title of the paper or by the name of an author, leading to a download, does not necessarily correspond with an essential contribution to the science of a potential author such that a reference is taken up in one of his/her forthcoming manuscripts.

- The perhaps best known number (ab)used as indication of the quality of a journal, is the Impact Factor (IF). The IF of a journal for year N is defined as the *number of citations* in year N to papers published in the journal in the years N-1 and N-2, divided by the *number of papers* published in the journal in the years N-1 and N-2. "The higher the impact factor the better"; but see what follows. Clearly this definition favours fields of science where papers have a "short life" and corresponding publication policies prevail. This is another indication that bibliometric numbers can have only meaning within a single field of science (and even there the publication and citation characteristics of subfields can be largely different).

- During my time as scientist in Delft (cf. Chaps. 8 and 10), the choice for a journal, that was considered to be the desired platform for a publication, was based on the personal quality assessment of usually the senior author(s). Factors that played a role are described by what I wrote above: "knowing the journal and being familiar with the type and quality of the papers published therein". This could lead to choice for a journal that perhaps was of somewhat lesser general standing, according to some measure, but that had a readership of likely more interest for the results to be presented. This changed during my time in Stuttgart. My Ph.D. students, and even more pronouncedly my post-docs, especially those from China, came forward with journal choices almost invariably based on the IF values: the higher the IF, the better. In some cases it took considerable effort from my side to convince my younger co-workers that another journal, that apparently had a smaller IF (I never looked these numbers up, but my co-workers were quick to inform me....) was yet better suited for the concerned contribution.

Moreover, IF values are sensitive to manipulation. Thus, as I, as author, experienced myself, the editor of a journal (in this case a journal of even distinct quality) can "advise" the authors of a submitted manuscript to refer to one or more papers published before in *specifically* the same journal (in the case referred to even papers were suggested which obviously did not fit at all to the topic of the submitted paper), with the obvious, sole intention to move upward, by these improper means, a measure for impact (as the IF) of his/her journal.

Within this context the following, at first sight likely unexpected, generally valid relation has been observed: *the higher the impact factor of a journal, the more likely that a paper in that particular journal is retracted.*[7] As interpretation one may suggest that an unhealthily large ambition of authors, to get a paper published in especially the top journals, plays a role here: thus immature or even fraudulent work can be submitted. At the same time this phenomenon then would expose the failure of the referees involved by the journal.[8] The last effect can be enlarged by the very short period of time assigned to the referees by the (top) journal concerned, in order to realize the speediest publication of a result thought to be startling and/or epoch-making (for example, see *"Cold Fusion; the Sun in a Pot of Water"* in Sect. 12.1 and *"Sheer Fraud; the Infamous Jan Hendrik Schön"* in Sect. 12.2).

To be the first journal publishing a result of such importance[9] means a lot for the reputation of the journal and enhances the number of citations to the journal and thus its IF, whether or not the work is later shown to be faulty or even a hoax. Although in the first place the authors submitting erroneous or deceitful work are to blame, the rush to publication by the top journals of results of supposedly expected extraordinary impact opens the door to loss of rigour in the evaluation of the manuscript as suggested by the observed apparent, positive correlation of IF and number of retracted papers.

---

[7] See: F.C. Fang and A. Casadevall, Editorial: *"Retracted Science and the Retraction Index"*, *Infection and Immunity*, (2011), 3855–3859.

[8] The referee does his/her task, voluntarily, unsalaried, in an honorary, anonymous capacity. Already therefore it is less appropriate in the *first* place to point an accusing finger at the referees: supposing a referee has no lack of time and is devoted to the evaluation of the manuscript to be reviewed, it may still be very difficult to impossible, without access to the raw data, to suspect, let alone identify, manipulation or data invention.

[9] This drive for priority for the journal differs from the drive for priority of the authors, as the latter priority is established by the date of submission of the manuscript to the journal, which is always printed in the published paper, usually at the beginning after the listing of the authors and their affiliations. Thus, in principle, the true originators will get all credits for being the first to have presented a specific scientific result, independent of the delay between the date of submission and the date of final publication of the paper. However, the historical record shows that such honesty is not always granted (by the scientific community); for such a case see *"The Rietveld Method; an Improper and Dishonest Namesake"* in Sect. 12.2, and see, in this context, also *"Diffraction line profile analysis; the Voigt function and the "Dream Team""* in Chap. 10.

Of course, I understood the background of the above mentioned drive of young scientists. They needed an impressive publication list for the next step in their careers. Also I myself was not insensitive to how I presented myself and my department to the world outside: a long publication list with papers in journals with the highest IFs may impress. Yet, as the preceding discussion indicates, the merit of the IF is highly overestimated for such a purpose. *There is no clear relation between the impact of a certain paper published in a certain journal and the IF of that journal* (also see the above cited paper by Marx).

Worse happens if the evaluation of the achievement of a scientist is also based on the IF values of the journals which published papers by this scientist. I have witnessed such unacceptable behaviour in also nomination committees for Professor positions; the candidates sometimes were even asked to include the IF value of each journal in their publication lists (also see a corresponding remark at the end of this chapter).

IJMR specific comments:

- Papers published in IJMR, and in materials science in general, do not generally acquire most of their citations in the first two years after the year of publication, but this is the time frame considered by the IF.
- Indeed the so-called Cited Half-Life for IJMR is larger than 10 years, distinctly larger than for well-known physics journals as Physical Review B.
- The average IJMR paper is cited distinctly more times than the average for journals in the field of materials science.
- Recognizing that most papers published in IJMR fall within the category metallurgy and metallurgical engineering, it follows that the IF of IJMR belongs to the top one third of all journals of this category.

The above statements may transfer a message that everything is allright with IJMR. However, doubts may be expressed. The above remarks are based on an evaluation made on the occasion of the centenary issue of IJMR in 2009. With reference to the development sketched above in this chapter, since 2009 a further dramatic increase of scientific literature has occurred, i.e. a dramatic increase of also submitted manuscripts. This development is not fully unrelated to the recent, enormous increase in the number of scientists of countries still developing a scientific tradition with civilized social norms, particularly in Asia, and submitting manuscripts. An Editor of a scientific journal is confronted with this paper "storm". The referees of IJMR and also the Editor cannot escape the impression that rather many of the submitted papers by authors of the implied nationalities were either of low quality and/or did not fit with the scope of the journal (authors did not even seem to be aware of the type of papers appearing in IJMR, suggesting haphazard selection of the journal; such a phenomenon was only seldom met in former years…). As a result the rejection rate of the manuscripts submitted to IJMR under my editorship increased to a little more than 80% in the last 5 years. This appears to be a phenomenon also experienced by other quality journals. As a result the number of papers published in IJMR remained more or less constant, which was unintended. Yet, even in this situation evidently the authors of the nationalities indicated above now are responsible for a very significant part of the work published in IJMR. This more or less represents a universal development in the scientific world and therefore cannot be criticized.

However, a problem has emerged that torments us till the present day. Of course, flagrant fraud and deceit have been of all times. However, one (at least every editor of a scientific journal, I presume) cannot escape the impression that the extent and scope of these crimes have pronouncedly increased in recent years. In my time as Editor-in-Chief of IJMR I was confronted with two cases of scientific fraud with respect to papers published in IJMR. As far as I know, in the more than 100 years of existence of IJMR such grave misconducts, leading to paper retractions *by the editor*, had not occurred before (i.e. in any case were not observed before). Also as a reader of the scientific literature one notes the now frequently reported cases of retractions of papers published somewhat earlier.

One of the most frequent reasons for retraction of a published paper by the editor of a journal is the finding that the same authors had published very or fully equal results elsewhere, of course and moreover without referencing the papers mutually. Also, using the same data as seemingly pertaining to different materials in different papers, or simply "inventing" data ("fabricated research") are no exceptions as well. The last fraudulent case I was confronted with as editor is described more fully in Sect. 12.2. In that case IJMR had to retract three papers; in total about 30 papers by the same author in various journals were found to be fake and were retracted; the suspicion exists that fraud also prevails in many other of the few hundreds of papers by this author.

Honesty requires admitting that the relatively largest amounts of fraudulent research papers stem from countries I indicated above as lacking establishment of a scientific tradition with its mores indispensable for the progress of science,[10,11,12]. This indeed holds for the two cases of fraud I met as editor, and as referred to above.

Most scientists and their papers provide only incremental steps forward in science. Only rarely great advances occur. On large and small scales in science, accepted theories/theoretical constructions (i.e. accepted by the very most of the scientists in the field concerned) usually define the framework for the scientific research that provides the incremental but significant steps forward in the understanding of nature. These frameworks, the context for the scientific work, can be called the "fundamental mindsets = paradigmas". Only if more and more reproducible results (from experiments) performed within in the framework of a prevailing paradigm in a certain field of science are in conflict with what is possible/predicted/allowed for within that paradigma, a hectic time in that field of science occurs, where conflicting proposals for the needed new theory are strongly debated, until a new paradigm, either totally different or, more often, a more or less modified, more comprehensive one, has become accepted and the former situation has been established, etc. These hectic, transition times of paradigm change may be called "revolutions in science".[13] Scientific progress evidently cannot be conceived as a process of ever-increasing knowledge that proceeds monotonously with time, even if the number of scientists would be constant as function of time.

Well known examples of such revolutionary transitions are the introduction of the relativity theories by Albert Einstein and the introduction of the quantum concept by Max Planck. These new theories/paradigmas are often not easily accepted. A notorious case is the theory of continental drift by Alfred Wegener. As everybody will see looking at the globe: South America and Africa may have been "glued together" in very ancient times. It may therefore surprise that the idea of Wegener of continental drift, published in 1915–1922, essentially based on very much more than this visual image, was

---

[10] For example, see: *"The Economy of Fraud in Academic Publishing in China" by Mini Gu (April 3, 2018)* at: https://wenr.wes.org/2018/04/the-economy-of-fraud-in-academic-publishing-in-china *and "Scientific Fraud in China" by Steven Novella (November 27, 2019)* at: https://sciencebasedmedicine.org/scientific-fraud-in-china/. See also footnote 4 in Sect. 12.2.

[11] For example, see: *"A shady market in scientific papers mars Iran's rise in science" by Richard Stone (September 14, 2016),* https://doi.org/10.1126/science.aah7297, at: https://www.sciencemag.org/news/2016/09/shady-market-scientific-papers-mars-iran-s-rise-science.

[12] For example, see: *H. Sabir, S. Kumbhare, A. Parate, R. Kumarand S. Das, "Scientific misconduct: a perspective from India", Medicin, Health Care and Philosophy, 18 (2015), 177–184.*

[13] In 1962 T.S. Kuhn published a book called *"The Structure of Scientific Revolutions"* (*The University of Chicago Press*), that I read as a young student about a decade later. This book is a milestone, which has influenced the discussion of the history of science in a fundamental way. The book has been criticized, but the basic idea has been largely accepted.

not accepted during his life time. After his death in 1930 the theory moved into oblivion. In particular experimental work after the Second World War confirmed the basis of Wegener's theory.[14] This example is recalled here to demonstrate how difficult it can be to present something new/revolutionary that does not fit in the ideas/theories of the established elite in a certain field of science. This is just human nature: those who are respected and have a high position and "live within" a certain paradigm, possibly based on own former work, are unwilling to become taught otherwise, as this is felt as humiliation associated with loss of status. Not always it is like that. But on much smaller scale, every scientist producing something new, deviating from consensus at the time in his/her field, may have a difficult time in persuading his/her peers. Examples of this are given in Chap. 9.

After having pointed out above the paper flooding, as caused by superficial, redundant, nonvaluable, "sliced" or even erroneous[15] papers, as a negative aspect of scientific "growth" in modern times, say beyond 1950, a concluding remark must be devoted to assessing the quality and future of science. A way must be found to stop the deluge of papers of present-day, in as far driven by status gain for a scientist by publishing as many as possible papers. To this end tools for quality assessment, at least not directly based on the number of papers published, have been proposed:

(i) One could count the number of references drawn by published work of some author. However, not all citations to a paper are positive citations… This approach can be refined by introducing certain factors which combine effects of "quantity" and "quality":

(ii) For example the number of publications of an author in journals with a so-called impact factor higher than a certain value can be counted (see above under "*Editor of a materials science journal*" for the notion impact factor and its criticism).

(iii) The so-called "h number", introduced by J.E. Hirsch in 2005, can be determined. It is given by the number of publications by an author which have been cited h times or more. Evidently this h number increases with age of an active scientist. One of more obvious disadvantages of this approach is that a scientist with few publications of which one or few more are very

---

[14] The story about the development and especially the difficult acceptance of Wegener's theory is very well described in a book by A. Hallam: "*A Revolution in the Earth Sciences, from Continental Drift to Plate Tectonics*", Clarendon Press, Oxford, 1973.

[15] Of course, science always corrects itself. Erroneous results do not survive. Sooner or later the true conclusions and data will prevail. But this can sometimes take a lot of time and a lot of additional work. In the meantime confusion and controversy can govern the debate. This obstructs the genuine progress of science. A lot is to be gained if errors by careless work by possibly moreover unqualified scientists can be avoided.

extensively cited, possibly suggesting correctly a high quality of this scientist and his/her work, yet gets a relatively low h number.[16]

This list can be extended easily……. Whatever one would like to propose, in any case "like must be compared with like": both publication and citation customs can differ strongly from field to field.[17]

It is often said, that, in the end, the evaluation of science and the scientist can only be done by other scientists who read and study the published papers; in case of a nomination possibly aided by personal knowledge of the scientist concerned (but this latter aspect can also bias the evaluation significantly….). This is the so-called peer evaluation. I have been member of numerous nomination committees, for assistant and associate professorships and full professorships. Only rarely published papers of the applicants were really read and studied. The first glance was on the list of publications and a quick judgement was reached, e.g. on the basis of the length of the list and checking if high quality journals were present. Only in a late phase, after many candidates had already been "sorted out", a focus occurred on the published work of the remaining few candidates, by really considering the science produced in more detail (sometimes even not that). There can be a lot of understanding for this approach, given the enormous amount of work involved in really finding out what a candidate has achieved and the committee members in truth do not have such amounts of time. However, this means that there is a lot to be said, especially in a first phase of evaluation of all candidates, in favour of judgements also aided by numerical tools of evaluation of the type as listed above (a combination of quantity and quality assessment), but the current, "one-dimensional" ones are too blunt (see footnote 17).

It may be hoped for that negative aspects of scientific "growth" eventually can be diminished, by the above discussed shift in quality evaluation. This will reduce the pressure on especially young scientists to publish too frequently too soon (see also Chap. 7). Thereby at least a bit of the publication tradition of former days is restored, where the average quality of

---

[16] Both the IF of a journal and the h-number of an author increase with time. This is caused by what can be called "citation inflation": (i) the number of citations in a paper have increased in the passed 50 years; (ii) one citation counts as one for a journal but counts as one for each of the co-authors of a cited paper and the number of authors of a paper has increased considerably in the last decades; (iii) the total number of papers has increased over time (see earlier in this chapter). As a consequence, if one would assign critical values to the IF and the h-number for, respectively, indication of high quality for a journal and excellence for an author (if that would be possible at all; see criticism in the main text), then these "critical" values must increase with time (S. Cranford, Matter 2 (2020), 1343–1347).

[17] For an interesting consideration, see S. Senn at http://www.senns.demon.co.uk/Bibliometrics%20in%20Mathematics%20and%20Statistics%20V5.htm.

a scientific paper was higher than nowadays and its long term "life", as illuminated by citations, was longer.

7

Academia, Academics and Academic Careers

Abstract A definition of the task of academic institutions is given. The intertwined nature of research and teaching in a single person; the model for the academic system proposed by Wilhelm von Humboldt. Consequences of the massification at universities are discussed. "Over-education" and "relaxation of standards". Increase of the distance between the Professor and the (Ph.D.) student. Administrative and fund raising tasks endanger severely the research and teaching performance; Professors more than rarely develop into sheer managers. The traditional academic career is obstructed by the nowadays lack of permanent positions; too much talent gets lost "on the way to the top".

The notion "academia" refers to academic organizations as universities and research institutes (e.g. the institutes of the Max Planck Society)[1] which have the goal: generation (by research) and preservation of basic knowledge (about nature), its application (in practical environments), and its transmission[2] to future generations. This description already indicates that there can be a close relation of two activities: research and teaching. Specifically, each staff member employed at an academic institution is normally, or ideally…, intensely occupied by both research and teaching/mentoring assignments. This *intertwined* nature of research and teaching/mentoring in a single person

---

[1] The academic system considered in this chapter can be in details different from country to country. The more specific remarks presented within this chapter are based on my experience obtained in The Netherlands and Germany and do not hold unrestrictedly for all countries in the world. The general tenor of the chapter is hopefully of relevance for every reader.

[2] At first sight it might be objected that this last aspect does not pertain to (e.g. Max Planck) research institutes, but as long as in these institutes a major (the dominant) part of the research occurs in the course of Ph.D. research projects, this transfer of knowledge takes place too (as from mentor (supervisor; doctor father) to Ph.D. student).

© The Author(s), under exclusive license to Springer Nature
Switzerland AG 2022
E. J. Mittemeijer, *How Science Runs*,
https://doi.org/10.1007/978-3-030-90095-3_7

is the essence of what happens at academic institutions. It culminates in the interaction of the Professor (both mentor and supervisor) and the Ph.D. student, where, in the ideal case, through the years of the Ph.D. project, the Ph.D. student develops from pupil to colleague (in science) of the Professor.

This unity of research and teaching is the model as expressed by Wilhelm von Humboldt in the beginning of the nineteenth century, which still governs (more or less) the philosophy of the university education in our time. At various places in this book I express that I am a strong proponent of this concept, because I have profoundly experienced myself the mutual fertilization of the partners in this interaction: for instance, the Professor who is confronted, for example during his/her lecture, with an only seemingly naive question by a student, opening, unintended by the student, a perspective the Professor had not observed before, and the student who is confronted with a scientist active at the frontiers of research having a wide overview of and deep insight into his/her field of science (e.g. see Chap. 13). Thus, unsurprisingly, I am not very much in favour of lecture courses at universities given by academics who only teach…, although for some basic courses (as in mathematics) in the beginning of a natural science study this may be an acceptable approach.

An "academic" then can be defined as someone having completed a scientific education (at a university), incorporating research (training), at a university or a research institute, leading to a Master of Science (M.Sc.) or a Master of Engineering (M.Eng.) degree (or similar degrees), or as culminating thereafter *for a part of these persons* in a Doctor of Philosophy (Ph.D.) degree.

The picture sketched in the first paragraphs of this chapter may always have been an ideal only partly realizable in reality, but it is for sure that nowadays, as compared to decades ago, we are very much more remote from this desired model. Self-governance of universities is impaired by the ever more explicit exertion of influence by governments, usually the principal financer (and by third parties; see also further in this Chapter). This is related to the enormous increase of the number of students (for example, from the sixties of past century till present-day (2021) with a factor of more than 10 in West-European countries, as Germany). There are a number of factors contributing to this effect, such as an increasing participation of students from working-class families (but still underrepresented, if the corresponding fraction of the population is normative) and of female students (now outnumbering male students, but not yet in "hard" science fields, as physics). At the same time at the universities no compensation in terms of a proportional increase in (permanent) academic staff positions was realized.

As a consequence the distance between and the interaction of a Professor and a student have increased and decreased, respectively.

Yet, I have aimed to establish this interaction in also the first years of the academic Materials Science study course for which I carried responsibility, e.g. by also giving a main, first year basic course myself and not leaving there the teaching/lecturing load on (only) the shoulders of assistants.

The "Humboldt" ideal is likely best realized during the course of a Ph.D. project. However, in view of the enormous increase in the number of Ph.D. projects in our times (see Footnote 2 in Chap. 6) one may wonder how realistic this expectation is. A Professor leading a large group will likely depend on his close associates for the daily guidance of his/her Ph.D. students. Then the Professor must take great care that there is enough intense interaction of him/her (the doctor father) with the Ph.D. students, so that he/she can genuinely be the leading personality in the research projects. This text immediately makes clear that there is a physical limit to the number of Ph.D. students with whom this Professor—Ph.D. student interaction can be realized effectively. In my experience the maximum number of Ph.D. students per Professor can be 10–12, where it is then taken for granted that in the group there are, say, three senior researchers who can take care of close (daily) guidance.

The above paragraph derives from the way I organized the research work in my department at the Max Planck Institute. I have observed (remote) colleagues with groups where the number of Ph.D. students was much larger. In those cases no research guidance of significance I suspect to be given by the Professor; the mentor role then is fully on the shoulders of the senior researcher involved in the same project as the Ph.D. student. This leads to questions about justified authorship of papers published from results originating from this project…

The role of a Professor is not only focused on research and lecturing/teaching; he has administrative and fund-raising duties as well. There is a certain danger that thereby a taste for more "management" tasks comes to the fore. This can lead to a possibly definitive move away from research (and teaching). Eventually the Professor may leave the laboratory/institute to become, for example, the director of a huge research facility or a large research organization, or the Rector of a university, or a high functionary at a ministry, etc. These examples I have all seen happen, more or less before my eyes (cf. Chap. 5). To avoid a wrong impression: I do not condemn these developments. Moreover, I am glad that high caliber persons can be found for these assignments. What I do disapprove is that Professors stay at their research and teaching position while effectively reducing their role to that of a manager and yet presenting themselves as a successful, active scientist.

The Professor title has a high societal status. The respect Professors, a (still) highly select group, experience reflects that they represent the apex of knowledge. Thus it does not surprise in a state of crisis (as in the recent Covid-19 virus pandemic) to see especially Professors as the advisors of the government and/or as participants in talk shows on radio and television and/or as operators of own podcasts.

Not especially the high societal rank of a Professor makes his/her job attractive. Indeed, to occupy a Chair at a university (one then speaks of a "full Professor"), and even more to be Director at a Max Planck Institute (see Chap. 13), has a lot of advantages (but note below the (nowadays) restrictions put between the brackets):

- Freedom in selecting the research area (follow one's passion; "do what you love"; but see the beginning of Chap. 11…).
- Transfer of one's knowledge to ((under)graduate and especially) Ph.D. students: Mentoring is a noble and rewarding (also for one's research) enterprise.
- Boss over one's time: what project worked on and when; no fixed working hours (but the amount of hours spent on work usually far exceeds that what is considered "normal" (i.e. 40 h per week or even less)).
- One is autonomous, or more correct: the superior (e.g. the Rector of the university[3] or the President of the research organization of which one's institute is a part) is "far away" (but one may be strongly dependent on the outcome of performance evaluations, e.g. by one's peers or by officers of funding organizations (see the last pages of Chap. 10)).
- Stability of position: Professors are tenured; they are civil servants in countries as Germany and The Netherlands (employed for a "life time"). Only in case of very severe wrongdoings they can be laid off (see Footnote 3).
- Acquisition of an international profile and cooperations.

In more recent years the enormous increase of the number of students has had negative consequences. Whereas the quality at the top institutions (e.g.

---

[3] Even the Rector may have little direct influence. My appointment as Professor at the Delft University of Technology in The Netherlands in 1985 was realized by the Queen: the Certificate of Appointment was signed by her (of course after advice by the university, and after screening of my person by Homeland Security…). Nowadays the situation has changed in the sense that the university appoints its Professors. Of course, in an indirect way a university can make the working conditions for a Professor that difficult that he/she resigns him/herself (for the description of such an exceptional case, see Footnote 8 in Chap. 3).

Max Planck Institutes and elite universities) may not have suffered, this needs not hold in general. The reduction of the staff/student ratio and a certain "relaxation of standards" (as in examinations) reflect a felt decrease in level of the incoming and outgoing (graduated) students. This is ascribed to the societal pressure both to increase the number of pupils completing a secondary school/gymnasium education, allowing university access, and to increase the number of students completing a university education,[4] which are quality affecting factors, as referred to at various places.[5] To be specific: it is a matter of course that those of our young ones capable of completing higher education should be offered this possibility, unconstrained by parameters as gender, or in which type of family one grew up, or financial conditions, or … However, already only the successful pressure exerted by those parents to let follow their child a gymnasium education also if the talent lacks, provides an indication that the effect described above is real (see the end of Chap. 3). But this is and remains a controversial issue.

From a sheer economic, so not cultural, point of view, it may also be doubted if our society needs so many people with a university education. According to such consideration one refers to "over-education". In recent years, especially in Germany, it has frequently been noted that a lack of professionals/skilled workers, having completed successfully a vocational training, is evident, which is ascribed to the drive within society to acquire a university degree, which has a (felt) higher status.[6] From this point of view, not everybody who has the talent for it, and certainly not those who are untalented in this respect, should follow a university education.

The attraction of a position at academia offering research and teaching/mentoring, as described above is felt by many having completed a Ph.D. project successfully. One may then decide to stay a few years at the same institute (provided the Professor offers this possibility) or at an other institute, possibly abroad. These positions are called post-doctoral positions.

---

[4] Funding of universities by the government can and has been based also on the number of students who successfully completed the academic education (also, the number of Ph.D. degrees granted has already been used in The Netherlands as a measure for a gratuity for the supervisor), which happens under the questionable supposition that this sheer mass criterion is indicative of quality.

[5] For example, see *B. Baldauf, "United Kingdom", in: Academic Careers: A Comparative Perspective (Eds.: J. Huisman and J. Bartelse), Twente, Center for Higher Education Policy Studies (CHEPS), 2001, pp. 17–34,* or *P. Swanzy, "Increasing HE enrolment and implications for quality", University World News, 5-10-2018* (https://www.universityworldnews.com/post.php?story=20181003101033477).

[6] This is for example expressed by a paper with a provocative title: *S. Hofert, "Werde Fachkraft - aber bloß nicht Akademiker" (="Become a professional—but just not an academic"), Der Spiegel, 6-3-2014 (* https://www.spiegel.de/karriere/fachkraeftemangel-kaum-akademiker-gesucht-a-956943.html*) (in German).*

One may decide so simply to satisfy the desire to extend one's academic experience for a couple of years and then look for a position outside academia (e.g. in industry). However, quite some of these post-docs foster the desire for an academic career.

The more or less traditional academic career may be described as follows: from Ph.D. student, post-doctoral researcher, assistant professor (lecturer), associate professor (senior lecturer, reader), to, finally, full Professor (= the "real" Professor, i.e. the Chair Holder). The names of the intermediate stages may change (as from country to country), but the sequence in principle is the same. In normal cases the Professor is tenured. The lower rank positions may be (largely) provided with time-limited contracts. This was different in former days: in 1978, after completion of my Ph.D. project, I was offered a position (in a group different from the one where I did my Ph.D. work) that was a permanent one. Although not many of such possibilities did exist, this was not a striking oddity in those days.

The phenomenon of temporary positions, with time slots of, say, one–three years, became prominent later. Actually the number of permanent positions at universities has decreased strongly in the last decades. One may wonder which driving forces are responsible for such a change.

The university may ideally still be considered as the place where knowledge is pursued for its own sake: "unravelling the secrets of nature". Obviously, currently this view is far too restricted. The massification indicated above has induced a tendency to diversify academic assignments such that different academics are assigned to, exclusively, either research or teaching or administrative (managing) tasks.

In one case, as an outspoken example of such diversification of academic assignments, that I witnessed myself, a university decided to appoint at the position of Dean of a large faculty an experienced, genuine manager from the industrial sector. The man had no scientific record of significance (not to speak of scientific excellence) that would allow lending him the title "Professor", but that was what happened upon his appointment as factual head of that faculty, and he used the title. This thereby also is an extreme case of deterioration of the value and meaning of the Professor title.

Academic performance is traditionally measured by the quality of someone's publications and by the opinion of his/her peers. However, these traditional academic criteria for excellence are nowadays extended by the success of an academic at funding organizations (i.e. in terms of finances), which, however, can have but need not have an, in any case unspecific, relation with the scientific quality of the reviewed academic. Moreover, the relevance of the science (to be) performed for practical applications has become an evaluation parameter as well (i.e. accommodating societal needs; in recent years a trend particularly strong in e.g. The Netherlands: at the time of writing this text the governmental financed, major science funding organization, NWO (Nederlandse Organisatie voor Wetenschappelijk Onderzoek (= Dutch Research Council)), provides twice as much funds for "strategic" research than for fundamental, "free" research[7]). By preferred funding of scientific projects with expected pronounced practical impact, governments in fact, distinctly and not necessarily positively, influence the course of science. It is clear that science projects originating from sheer curiosity (see above: "knowledge pursued for its own sake"), which often are of high risk character and of long-term nature, then have lesser chance. It appears that these developments are unstoppable for the moment.

It is not by chance that many study courses at universities have been developed in recent years which, at least at first sight, have only a slight scientific connection. For example it is possible to get an academic degree in "tourism". There is no doubt that this can be useful for society, but the students following such study courses and the academic staff involved do hardly contribute to the progress of science in the sense of Chaps. 4 and 6.

Universities may have no chance to counteract the tendencies hinted at above, recognizing that they have experienced increasing financial constraints and are also judged as a whole by society on the basis of parameters described above. As a result more top-down management within the university has emerged: the "primus inter pares" model is affected. The priorities of the university research and teaching programs are defined by the "university management". The self-governance of time and funds becomes reduced in a general way. In extreme cases the managers in the organization may even

---

[7] Vannevar Bush was the director of the Office of Scientific Research and Development in the USA and as such effectively coordinating the USA (military) research and development of the USA during the 2nd World War (involving also the Manhattan (atomic bomb) project, that he initiated; see also text under "*Abdul Qadeer Khan; Scientist, Spy and Hero; a Father of the Atomic Bomb*" in Chap. 11). He was of the opinion that science should be directed by the scientists themselves. In his report to the President, entitled "*Science The Endless Frontier*" (July 1945), he wrote: "The publicly and privately supported colleges, universities, and research institutes are the centers of basic research. They are the wellsprings of knowledge and understanding. As long as they are vigorous and healthy and *their scientists are free to pursue the truth wherever it may lead*, there will be a flow of new scientific knowledge to those who can apply it to practical problems in Government, in industry, or elsewhere." (italic by me). The governmental financed National Science Foundation, the dominant science funding organization in the USA, in fact has followed this philosophy until recently. However, also in the USA the tendency to steer research top-down, such that it serves (supposedly) practical needs for the country, is becoming more and more prominent, as expressed by the recent "Innovation and Competition Act" that indicates "key technology focus areas", as advanced materials science, biotechnology and artificial intelligence, which will be provided with specific funding for research.

introduce time sheets documenting the presence of the researchers at the laboratory/institute. For example, this happened in the Laboratory where I was employed as a young scientist. But, I am glad to say, only for some time: the nature of science is only partly compatible with bodily presence of the scientist at his/her official working place.

It is this type of developments that also has led to the introduction of more and more time-limited contracts for the academic staff at universities. Thereby a more immediate response to the whether or not availability of required budgets can be realized. Moreover, in this way also not well functioning groups can be indirectly "punished" and flourishing groups can be indirectly "rewarded". This last aspect is not a bad thing as dormant groups did occur, especially in the past. But, according to my own observations, such groups are nowadays more of an exception, have "died out". All now participate in "the struggle for survival".

One probable outcome of the sketched developments is the concentration of genuine cutting-edge research (and teaching/mentoring) in a few elite universities (e.g. Harvard University and the Massachusetts Institute of Technology in the USA; it is no coincidence that these universities are financed privately to a significant extent) and elite research institutes (as those of the Max Planck Society in Germany).

The consequences of all this for young scientists wishing to follow an academic career are severe. In relation to the enormously large number of post-docs, assistant professors, etc., all with time-limited contracts, very few permanent positions, especially Chairs, are available. Those desiring a Professor position feel an enormous pressure, having to produce an impressive series of papers, showing great skill in acquiring funds and performing science on a "relevant" topic.

A nasty consequence of the scarcity of permanent academic positions is the phenomenon of "post-doc hopping": the young scientist moves from one position to another (often at another university/institute), becomes older in the meantime and loosens also thereby "attractiveness" for the rarely available vacancy of a tenured position. It may become too late to arrive at the insight that a permanent academic position is unattainable and that one has to look for a job elsewhere in society. This can also happen with good scientists, as I could observe myself from nearby. One may feel obliged to employ such a scientist, who was considered elsewhere as too old (as for a vacant Chair), on again a temporary basis as no other possibility exists, to avoid an undeserved, disastrous end of a career.

Indeed, the absence of a long-term prospect and the lack of job stability have been indicated as causes for *not* choosing an academic career at a university.[8]

Of course, it is argued that flexibility, job/position mobility can be expected of young scientists wishing to embark on an academic career path. Moreover, a permanent employment in an early stage may lead to loss of driving force and fixation of the area of science and competence, in the absence of gone astray eagerness for change and new fields of research. I have heard that often in the seventies and it certainly applied to a small part of the academic staff (tenured) at the university where I was a young Ph.D. student at the time. However, the currently prevailing scarcity of tenured positions, which is reflected in the severe selection of scientists who will be offered corresponding contracts, has made such underperformance on the few permanent positions practically unlikely in general.

During my time in Stuttgart, in the second half of my career, the lack of tenured positions for young promising scientists has led a few of my most promising, both male and female, young Ph.D. students of German origin not to go for a post-doctoral position and try to climb the academic ladder. The reason was, as they explained to me, that economic stability at their stage of life (often as part of a couple with (forthcoming) children) was considered of crucial importance. The salary as post-doc/assistant professor is modest (this generally holds for all scientific academic positions) and significantly smaller than what is offered elsewhere at comparable levels of competence (e.g. in industry) and, moreover, non-academic positions can be more stable/permanent. In combination with the very high uncertainty to acquire in the end a full Professorship, it may be understandable why so much scientific talent gets lost so soon.

Nonetheless, especially most of my Chinese, and also Indian, Ph.D. students and post-docs decided for an academic career. Quite a lot of them later acquired a Chair in their homeland (see end of Chap. 13). Evidently, their attractiveness for a Chair in China was relatively large after their built-up experience and excellent performance at the Max Planck Institute. Further, their chances were relatively improved in view of the relatively large number of available Chairs at the time in China for returning scientists. Yet, also their path was not sure to lead to the desired goal; the competition they were subjected to was very strong as well.

---

[8] For example, see: *R. S'Jegers, J. Braeckman, L. Smit and T. Speelman: "Perspectieven uitgestroomde wetenschappers op de arbeidsmarkt" (= "Perspectives graduated scientists on the labour market"), Vlaamse Raad voor Wetenschapsbeleid (= Flemish Science Policy Council), 2002 (in Dutch).*

Although I can understand and respect the reasoning of those Ph.D. students who decided against an academic career, and indeed the chance for a Chair in Germany/West-Europe is very small, I do regret their apparent lack of some necessary daring. For those Ph.D. students I supervised in Delft, in the first half of my career, and who later acquired a Chair (see end of Chap. 10), I was used to more willingness to move elsewhere and try one's luck. The outcome of their academic route was very unsure as well, but they persevered and eventually succeeded.

Notwithstanding the above modifying remarks derived from my personal experience, the conclusion of the above consideration can only be that permanence of an academic position must be given earlier than nowadays in the academic career; i.e. we need more permanent positions below the rank of the Chair Holder. Severe selection at that stage then must assure that such a permanent "assistant professor/"research fellow" position is only given to those strongly talented and subjected to the inner drive of the genuine scientist considering his/her research and teaching/mentoring assignments as a calling. These are big words. I believe them to be in order here. If these boundary conditions are satisfied then also the functioning of the scientist remains at high level if eventually an appointment as Professor does *not* happen.

8

Immersed in Materials Science

Abstract A brief history of the Laboratory of Metallurgy/Materials Science. Peculiarities of living in my house in Delft. What is Materials Science. Becoming a genuine materials scientist. The interaction of iron and nitrogen as a research theme for a lifetime. Developing a small research group; "reanimating" co-workers of the Laboratory. Interaction with industry as a rewarding experience for a fundamental scientist. The Linguae francae of science. My first lecture in German; a special experience. A memory of Professor Meijering. Encounters with science and scientists behind the "Iron Curtain" in Poland and the DDR. How I got my grand piano.

The Laboratory of Metallurgy/Materials Science

Already before the Second World War, metallurgy was established as an obligatory course for a number of faculties (including the physics faculty!) at the Delft University of Technology (TUD). After the Second World War, also in view of the growing importance of metallurgy and the increase in the number of students, ideas about a separate laboratory for metallurgy at the TUD emerged. The initiative was supported, if not also carried, by the Philips Research Laboratories (NatLab; see under "*From Science to Management; the Career of Professor Okkerse*" and footnote 18 in Chap. 5). There has always been a strong, albeit informal link with "Philips", as a substantial number of the Professors then and later appointed on Chairs in "metallurgy"/"materials science" had come from Philips to the TUD (Burgers and his successor Okkerse are only two examples). The idea behind the new laboratory was to combine the more engineering side of metallurgy with the more fundamental side, in order to create a fruitful symbiosis. The fundamental metallurgical

E. J. Mittemeijer, *How Science Runs*,
https://doi.org/10.1007/978-3-030-90095-3_8

91

research was represented by Professor Burgers, on a Chair of the Faculty of Chemistry, and Professor Druyvesteyn, on a Chair of the Faculty of Physics. The engineering/technical research, devoted to topics as casting (solidification), welding and soldering and mechanical properties, was represented by initially two professors appointed on Chairs in engineering faculties.

It would take quite a number of years before the new Laboratory of Metallurgy could be opened: that time had come in 1961. So when I moved to the Laboratory of Metallurgy for my graduation project in 1970, this building was still rather new and, in the meantime, Burgers and Druyvesteyn had been succeeded by Okkerse and Penning (see Chap. 5).

More or less parallel with the above development the wish arose for an independent academic education to "metaalkundig ingenieur" (we would now say: M.Eng. (=Master of Engineering) or M.Sc. (Master of Science) in metallurgy).[1] Such a complete academic course was realized upon founding in 1952 of the "Tussenafdeling der Metaalkunde" (possibly best translated as: Interdisciplinary Faculty of Metallurgy), i.e. a faculty in-between other faculties. The multidisciplinary Laboratory of Metallurgy carried and was core and heart of this academic course on metallurgy. And of course it was possible for students of other faculties to perform, after their first years, their graduation project here: e.g. chemistry students in the group of Prof. Burgers and physics students in the group of Prof. Druyvesteyn.

The Laboratory stood at the corner of the Rotterdamseweg and the Jaffalaan, with its long side parallel to the Rotterdamseweg. It consisted of two wings: a main wing adjacent at the Rotterdamseweg (shown in Fig. 8.1) and a parallel wing comprising especially the "Halls", where large and heavy equipment was installed. The second, top floor in the main wing accommodated the Burgers and Druyvesteyn groups.

On the other side of the Rotterdamsweg and opposing the Laboratory was (and still is) the porcelain factory "De Porceleyne Fles", trading as "Royal Blue". This is the famous producer of "Delft Blue" earthenware. This manufacturing company was founded in the 17th century, and is the only surviving of such factories from that time. A tableau of original, old Delft blue tiles, each showing a child performing a different play, as played by children in the 17th century, was mounted as decoration next to the fireplace in the living

---

[1] The nowadays abolished academic degree "ingenieur" is abbreviated with "Ir." in the Netherlands. Such a degree does and did not exist in Anglo-Saxon countries and many other countries. In Germany (and Austria and Switzerland) the similar degree was "Dipl.-Ing." (= Diplomingenieur) and, for example also in Poland this degree was granted. However with the "Bologna reform", since 1999 (see footnote 16 in Chap. 13), these degrees in the European Union were eventually all replaced by M.Eng. or M.Sc..

Fig. 8.1 The main wing of the Laboratory of Metallurgy (later named Laboratory of Materials Science) situated parallel to the Rotterdamseweg in Delft (taken from *P. Jongenburger, A.J. Zuithoff, M.J. Druyvesteyn and W.G. Burgers, "Technische Hogeschool Delft—Laboratorium voor Metaalkunde", 1961*). The second, top floor in the main wing accommodated the Burgers (later Mittemeijer) group and the Druyvesteyn (later van den Beukel) group

room of my old house in Delft (see below this intermezzo). Today "De Porceleyne Fles" is a great tourist attraction: of the order of 150.000 visitors come each year.

I have to admit that during all my more than 25 years in Delft I have never visited "De Porceleyne Fles". However, as some sort of compensation, from my office at the second floor of the Laboratory (the former Burgers office) I could well see the top, second floor of "De Porceleyne Fles". Thus I had a very good look at the nice girls sitting there and delicately painting the thereby to be decorated vases and other ceramic products, which were so wanted for by the tourists.

As elsewhere in the world, other material classes became important in the course of time and thus also in Delft the original metallurgy course was broadened in 1981 to a materials science course, incorporating ample attention for also material classes as polymers and ceramics, which meant that the bond with other faculties, notably chemistry, became only stronger. At some stage also the name of the Laboratory of Metallurgy was changed into Laboratory of Materials Science.

It appeared that it was very difficult to build the "bridge" between the engineering groups and the fundamental groups in the Laboratory. Professor van den Beukel, the successor of Penning, and my closest colleague during the time I occupied Burgers Chair (see Chap. 10), once said to me, with a little sarcasm: *"It was really something like a revelation, a miracle, that suddenly in the discussion of a lecture, Prof. so and so,* (he mentioned the name of the professor on welding and soldering technology) *suddenly used the notion "dislocation". It felt like a shock."* (regarding the concept "dislocation", see footnote 8 in this chapter). Thereby one factor contributing to the eventual decline of the Laboratory of Materials Science may have been indirectly suggested: whereas the two fundamental groups reached high international standing, this could not be said for the engineering groups in general (i.e. with temporary, person connected exceptions). But decisive was the lack of students:

Delft was the only place in The Netherlands where an academic materials science education was offered. It then surprises to observe that, in too many years, it appeared rather difficult to attract students for this academic education. And that although industry generally had strong interest for graduated materials scientists; non-employment of the graduated students was extremely rare. There were good years with 30–40 freshmen, but it also happened that the number of first year students was about five. Certainly this had at least partly to do with being it difficult for pupils at a secondary school (HBS or gymnasium) to obtain a formative picture of materials science, in contrast with chemistry and physics which were and are taught at secondary schools *under these names.* Yet, in Germany materials science courses attract, also relatively, significantly more students, as I experienced myself in Stuttgart. Perhaps this difference has something to do with Germany being a more industry minded country with a longer such history than The Netherlands characterized by a dominating mercantile culture.

After I had left for Stuttgart in 1997, rather meagre years happened in Delft with a very strong low in 1998: only three first year students. In 2001 the bachelor part of the materials science course was closed. In 2006 the Laboratory of Materials Science was demolished… Materials science has survived at the TUD on a smaller scale as a pure Master course and has been incorporated both into the building and into the organization structure of an engineering faculty.

My House in Delft

Some time after the decision to stay in Delft, having accepted the position in Korevaar's group, I bought an old, relatively small townhouse at the northern edge of the city centre. The house stood at a square, the Paardenmarkt (=horse market), that was not used any longer as a place to trade horses, but was used as a car parking place for the local residents, which spoiled considerably the attractiveness of the place.

The Paardenmarkt had not been the result of deliberate planning by a city board from the past, but it was "simply" a result of the devastating explosion of the gun-powder house in 1654.... Renovation works in front of my house during the time I lived there revealed parts of thick walls, remnants of that distant past. The house had its charm (it was partly put under monument preservation), but, unfortunately, many defects regularly made living there less pleasant:

The level of the groundwater could sometimes be very high, causing the cellar to be filled up to considerable extent with water. This became a serious problem, especially if the pump had not automatically become active, which was often the case. Then I had to empty the cellar, bucket after bucket, as the wetted pump did not dry soon (a large part of Delft, including my house, is located below sea level; the local waterboard, responsible for the management of the water (level), dates from the 13th century and is nowadays called "Hoogheemraadschap Delfland"; a special tax is levied to finance its task). The water rising in the cellar could have attained a level such high that heating of the boiler and the central heating system, both located in the cellar, was no longer possible, as the humidity kept extinguished the pilot light; which obviously was uncomfortable in winter. There was no way to suppress the generally high humidity level in the house; especially the wall between the kitchen and the living room deteriorated gradually by upward move of humidity.

Moreover, the former renovator of the house had put a roof on the house that, located next to the neighbouring, higher house with a perpendicular side wall, was partly inclined. The thus created small and sharp, asymmetric wedge caused the wind, especially during storms, not uncommon so close to the coast as Delft is, to become very turbulent there, causing the tiles of the roof to move about, opening the roof and thus leading to pronounced leakage of water, from the storm accompanying rain shower, through the roof. I had to climb the roof often to repair the arrangement of the tiles, and to remove the excrements of cats from the neighbourhood, which seemed to favour the gutter between the two houses, at the bottom sharp edge of the wedge, as the preferred site to relieve themselves.

A main water-supply pipe ran parallel to the outer wall, at its inner side, at the front of the house. During one very cold winter the water in the pipe froze, and the associated expansion did fracture the pipe, causing, upon subsequent thaw, flooding in the entrance hall. Also the sewage system was very old, as I came to learn to my annoyance: once I came down in the early morning from my sleeping room to observe that the system had operated in the opposite way by depositing its contents not only in the toilet but, upon overflowing it, also on the floor around. The stench was unbearable. A radical restructuring was necessary implying opening of the tiled bottom in lavatory and entrance hall, and also involving replacement of no longer allowed for lead tubes installed in the house in a long ago past.

The above may serve to indicate why I decided that I would never ever buy or live in an old house again.

Yet, I lived in the house at the Paardenmarkt for about 18 years. The greatest attraction of the house certainly was its location at the edge of but still within the old city centre. Everything essential for daily life could be acquired and arranged for by walking and the atmosphere of old Delft was indeed very enjoyable, not to be compared with the neighbourhood of the Roland Holstlaan, where I had lived at three different addresses before. And there was the cat of my neighbour, called Caesar, who visited me often in the evening by scratching with a paw the glass of the window to the garden behind which I sat working at the table; I loved the cat.

Further, I could go by bike to the Laboratory, a trip of about 10 min. I enjoyed that, cycling along the Oostplantsoen, the Oosterstraat and the Oranje Plantage to the Oostpoort. There, more or less at the end of the Oranje Plantage, stood the house where Antonie van Leeuwenhoek was born. Van Leeuwenhoek is probably the greatest experimental scientist to be associated with Delft.[2] He lived all his life in Delft. It was only during the preparation of my Farewell Lecture in 1998 that I "discovered" that I had been cycling 18 years passing the location of van Leeuwenhoek's place of birth without ever realizing that.

Upon arriving at the Laboratory of Metallurgy I dropped my bike either in the bicycle storage before the main entrance or, obstructing a possible theft, in the one in the basement. Then I went to my office. During my time as a Professor, my office (the former Burgers' office; see Chap. 5) was in the

---

[2] Antonie van Leeuwenhoek (1632–1723) was an "autodidact", a self-taught man and a "man of means" (cf. the remarks about Charles Janet in the intermezzo *Looking at the Periodic System….*", in Chap. 5). By a special technique, that he had developed himself, he was able to produce lenses, in fact small glass spheres of high quality, which allowed him, by application of the corresponding single lens microscopes, to attain resolutions unsurpassed for the time he lived in. His observations were of revolutionary character. He never wrote a book or published papers; his findings are presented in very many letters to the Royal Society in London. He was the first to observe single celled organisms (bacteria, protozoa). Remarkably he thus was also the first to observe spermatozoa. In retrospection one can consider him as a founder of a field called microbiology. Van Leeuwenhoek has a special meaning for me, not only because we both had a strong connection with Delft, but in particular because of my own interest in light and electron microscopy, a topic that I lectured about for many years in a course called "Diffraction and Image Formation".

second floor. I normally took the stairs. The Laboratory of Metallurgy had a first and a second floor, but was thus constructed that a third floor could possibly later be built on top (one had thought that such enlargement would become necessary. It never came to that; see the preceding intermezzo). This optimistic foresight brought about a construction of a main stairwell where the stair case to the hypothetical third floor was already partly constructed: it ended very close to, in fact practically at the roof of the building. Being more than once in "deep" thought, I went up the stairs, did not enter the second floor but continued climbing the stairs, with the result that I knocked unpleasantly with my head at the roof of the building and "woke up". Indeed this concentration, more or less forgetting my surroundings, also happened not very rarely upon cycling and could let end me up at a place where I had not planned to go.

Materials Science

"Materials science" is a discipline at a position intermediate between chemistry, physics and engineering.[3] It evolved since the middle of the past century, with metallurgy at its core. Experience had taught that teams composed from scientists of such various disciplines, as indicated here, were a prerequisite to arrive at fundamental understanding of the behaviour of materials in practice. This word "practice" immediately underlines that a material must have an importance beyond its mere existence:

A material is a substance with a present or a realistically expected future application for mankind[4]

---

[3] This *Intermezzo* to a significant extent derives from Chap. 1 of my book "*Fundamentals of Materials Science*" (*Second Edition, Springer, 2021*), where more background is given.

[4] I have (now; cf. the reference in footnote 3) added "realistically expected" in this definition, because I have noticed that often the most exotic substances, certainly with specific, sometimes scientifically interesting properties, are denoted as "materials", although no materials scientist or materials engineer would foresee any application in a not too distant future. This untrue and misleading denomination is used, by especially (solid state) physicists and chemists, to apply for (government or European) funds available for the development of materials or in an effort to comply with a popular, reigning trend in the science-support policy of the society/the government (cf. the hausse in the abuse of the word "nano*materials*"). Thus I have observed that "suddenly" a faculty of chemistry had "found out" that a very large part of its research concerned "materials", could be decorated with the notion "materials science", and had designated itself accordingly in its strategy papers and fund applications. When asked privately and not in this context, these physicists and chemists do not at all want to be considered as materials scientists, but they see themselves as "pure" physicists and chemists, moreover typically considering engineering aspects as part of materials science far beyond their interest.

Using this definition one can distinguish between *natural* materials and *man-made* materials. Wood is such a natural material as well as some metals as found in elementary form (copper, gold). Steel is man-made as well as plastic.

In the course of time, departing from metals, other material classes were found to be subject to the same interdisciplinary approach as metals, in order to make the connection between the material and its behaviour in practical applications. Thus material classes as ceramics, polymers, semiconductors can be mentioned. Some of us, in recent days, also consider composites, biomaterials, carbon nanomaterials, as graphene, etc. etc., as separate material classes. It does not make much sense to start a debate about the sense of defining all these types of materials as separate material classes: the boundaries between the material classes are in any case nebulous; it is already rather difficult to impossible to provide a really watertight definition of what a metal is.....

The signifying, characteristic feature of materials science is the study of the relation between a material and its properties, by analyzing the structural hierarchy in a material, which is usually called the *microstructure*. Its study and effect on the material properties is THE distinguishing feature of materials science as compared to merely solid state physics or solid state chemistry. The materials scientist bridges various length scales: On the *nanoscale* (nm range; 1 nm $= 10^{-9}$ m) an atomistic approach is required. The *microstructure* scale, in narrower sense, covers the nm to mm range. Here the defects in the (atomic) structure, dislocations, grain boundaries, stresses, etc., strongly influence to even govern the material properties. Finally the *macroscale*, from the mm range to far beyond, concerns the action of material components in engineering constructions, where, for certain properties the material can be considered as a "continuum". Going upward from one length scale to the subsequent one, the knowledge of the deeper levels is required to be able ultimately to understand and describe the material behaviour. In other words: the endeavour of the materials scientist involves the "transition from local (individual atom/defect) to non-local (polycrystalline, macroscopic) descriptions".

In Chap. 5 it was remarked: *"the borders between disciplines as chemistry and physics do not really occur; they are at best very vague. The delineations of a scientific discipline, which non-scientists and beginning students believe to exist, evaporate upon one more and more delves into science."* Against this background one may question the meaningfulness of trying to define materials science, after all and moreover an interdisciplinary activity, as a separate discipline. This may be a justified and at least understandable remark. However, at very many universities worldwide, faculties, departments, study courses of materials science (and engineering) exist. In my eyes, next to the above discussion, already that suffices, for all practical purposes (!), to accept materials science as a separate scientific discipline.

My first office during my time in Korevaar's group was at the side of the Laboratory where the large halls were concentrated. These halls contained mainly big equipment for mechanical testing, casting etc. On the first floor in this part of the Laboratory there was a row of offices; one of these was used by me. Here, departing from the field of physical chemistry, I started to become a genuine materials scientist.

DAF is a well-known Dutch company producing high quality trucks mainly sold in Europe but also to countries outside this continent. The engine contains a crankshaft. In a running engine crankshafts are subjected to a type of cyclic loading that is called "rotating bending fatigue". Fatigue is the loss of strength occurring upon cyclic loading at a stress level below the one that would cause immediate failure. During such fatigue loading, changes in the microstructure of the material occur, usually invisible from the outside, which ultimately yet lead to fracture. The loading of the crankshaft mentioned here implies the imposition of stress cycles, having a compressive part and a tensile part, especially in the surface region of the material component. To counteract the negative effects of the loading stress, the assignment for the materials scientist is to find ways to strengthen the material (especially in the surface region of the component). At this point my now direct involvement with materials science got its perhaps most practical manifestation.

Strengthening of the material of a component in its surface region is nowadays called "Surface Engineering". In the present case (the crankshaft, made of steel) the microstructure can be changed and strengthened by precipitation of tiny particles within the material and/or by application of a process that invokes an internal stress in the surface region counteracting an external, tensile loading stress. Such a possible process is "nitriding", implying the introduction of nitrogen into the steel. "Nitriding" would become and remain a fundamental research theme of importance during the rest of my scientific career, albeit it moved me rather far away from the above described origin of these activities; in fact this career-long project can better be indicated as the investigation of the interaction of nitrogen and iron (and iron-based materials).

Nitriding

The thermochemical process denoted with the name "Nitriding" has been the most versatile and efficacious method of "Surface Engineering" of iron-based materials (as steels) for already many decades. It involves the introduction of atomic nitrogen (N) into the surface region of the component. The process also today finds a still widening field of applications. Its tuned operation allows drastically pronounced improvement of the fatigue resistance, the wear resistance and the corrosion resistance.

Various ways for introducing nitrogen into the material are possible. The classical one involves the dissociation of (gaseous) ammonia (NH_3) at the surface of the material, leading to atomic nitrogen dissolved in the substrate ([N]) and hydrogen gas escaping from the surface (H_2):

$$NH_3 \leftrightarrow [N] + \frac{3}{2}H_2.$$

The nitrogen can react with iron and give rise to the development of a layer of iron nitrides at the surface of the material, called *compound layer*. Underneath this compound layer, the inward diffusing nitrogen, in the so-called *diffusion zone*, can react with alloying elements dissolved in the iron-based substrate, as chromium (Cr) and aluminium (Al), leading to the development of very tiny particles of alloying element nitride, as chromium nitride (CrN) and aluminium nitride (AlN).[5] The compound layer can bring about distinctly enhanced tribological and anti-corrosion properties. The diffusion zone is responsible for the large improvement of the fatigue resistance (see Fig. 8.2).

The development of the tiny alloying element nitrides in the diffusion zone induces a tendency to (lateral) expansion of the nitrided zone. However, this strived for expansion is counteracted by the core of the component and the desired local lateral expansion cannot occur. As a result a compressive, residual,[6] internal stress occurs in the diffusion zone (see Fig. 8.3). Upon fatigue the component is subjected to many cycles of an external, within one cycle usually subsequently compressive and tensile, loading stress. Fatigue can ultimately lead to crack formation and crack growth at the surface, under the

---

[5] These nitrides often appear as platelets of lateral size a few tens of nm (1 nm = 10^{-9} m) and a thickness of only a few (e.g. five) mono(atomic) layers, i.e. the order of one nanometer. They cannot be seen under a light optical microscope; a (transmission) electron microscope must be applied.

[6] *Residual* stresses are internal stresses, acting within a material component in the absence of any external loading stress. They can result after the material has been subjected to some treatment, i.e. here the nitriding treatment causing the precipitation of miniscule nitrides within the diffusion zone (see Fig. 8.3).

The Nitrided Surface Region

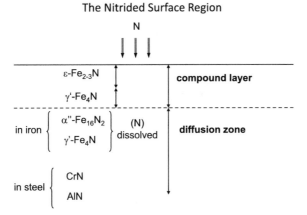

Fig. 8.2 The nitrided surface region of iron or steel components. Nitrogen (originating from e.g. ammonia in the outer gas atmosphere, that dissociates at the surface of the iron/steel component) reacts at elevated temperature, say, 550–590 °C, with the iron of the component/workpiece. Adjacent to the surface of the component the so-called *compound layer* develops, usually dominantly composed of two iron nitrides, γ' and ε. Underneath the compound layer, the inwardly diffusing nitrogen gives rise to development of the so-called *diffusion zone*: (i) in case of a pure iron component, the dissolved nitrogen can precipitate upon cooling to room temperature as α'' and γ' iron-nitride precipitates; (ii) in case of an alloyed steel component, the inwardly diffusing nitrogen can react with alloying elements dissolved in the substrate, as chromium (Cr) and aluminium (Al), leading to the development of tiny alloying element nitride precipitates, as chromium nitride (CrN) and aluminium nitride (AlN). The *compound layer* can be responsible for improved wear and anti-corrosion properties. The *diffusion zone* contributes to enhanced fatigue resistance

influence of the *tensile* loading stress part of the compressive-tensile stress cycle, in the surface region of the component. If a *residual, compressive* stress is present in the surface region, then this stress component counteracts the effect of the tensile loading stress and thereby the development and growth of a crack is hindered (Fig. 8.4). After nitriding the component can bear a much higher load (more than 100% higher is no exception) until failure upon fatigue is induced.

The above text can be considered as a too crude introduction to "nitriding", but more or less well represents the state of knowledge at the start of my involvement with "nitriding" in 1978. As a matter of fact, at the time most knowledge was based on phenomenology and largely derived from innovations in industry. An overview of current understanding is provided in: *E.J. Mittemeijer, ASM Handbook, Volume 4A, 2013, pp. 619–646.*

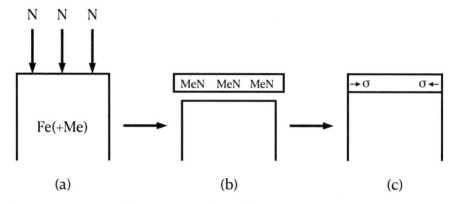

Fig. 8.3 **a** Upon nitriding tiny particles (called precipitates) of alloying element nitrides (indicated as MeN; Me can, for example, be chromium (Cr) or aluminium (Al)) develop in the diffusion zone (see Fig. 8.2). **b** This leads to the tendency of (lateral) expansion of the nitrided zone. This strived for lateral expansion is counteracted by the core of the component, because the nitrided zone is bonded to the unnitrided core of the component. Therefore the desired lateral expansion, shown in **b**, in the nitrided surface region, the diffusion zone, cannot occur. **c** As a result a compressive, residual ("residual" means: remaining after the material has been subjected to some treatment (here: nitriding)), internal stress, indicated by the symbol σ, occurs in the diffusion zone (in this discussion the presence of a compound layer (see Fig. 8.2) is ignored)

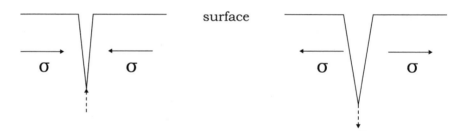

Fig. 8.4 Effect of (residual, internal) stress, σ, parallel to the surface of a piece of material (component) on the propagation/growth of cracks perpendicular to the surface. Whereas a compressive stress promotes closure of the crack/retards further crack growth, as indicated in the left part of the figure, a tensile stress promotes crack growth, as indicated in the right part of the figure

Start of Nitriding Research

Researchers at DAF had established a working group from scientists and engineers at Philips, the Delft and Eindhoven Universities of Technology (TUD and TUE, respectively) and DAF. The "Delft" involvement implied the participation of Korevaar, who had proposed, upon my arrival in his group, that I would take on me this assignment. I quickly saw that, to acquire insight into the effects of nitriding, fundamental research with model specimens was required. There existed a vast amount of publications in the technical literature about nitriding, but these were either more or less reports of experiences in industrial practice and/or focused on the nitriding behaviour of steels containing many elements and usually possessing a complicated microstructure, which formed an obstacle to arrive at definitive conclusions. Thereby gaining understanding on a fundamental level was impeded.

So, I decided to start a nitriding project satisfying two conditions: (i) the chemical potential of nitrogen[7] in the "nitriding atmosphere" must be known and controllable and (ii) nitriding experiments would start with pure iron and a few well-chosen binary iron-based alloys as iron-chromium (Fe–Cr) and iron-aluminium (Fe-Al). Regarding condition (i) only a gas atmosphere, composed of ammonia (NH_3) and hydrogen (H_2), allows such control of the chemical potential of nitrogen by fixing the ratio of both gases and realizing a sufficiently high gas flow through the furnace so that any decomposition of ammonia could be neglected. Perhaps surprising to an outsider, complying with the second condition was not trivial. To obtain really pure iron was (and still is) not easy. Here we had luck, since Prof. Meijering (see further below) had obtained a large sample of such highly pure iron from his contact with the Philips Research Laboratories (NatLab; where he had been employed before) and we were allowed to use this material. The binary alloys were produced in our own laboratory by alloying in a controlled way pieces from this sample of highly pure iron with specified amounts of highly pure alloying elements. As I learned, also this was not a task that could be performed in routine fashion.

My very able assistant Hans van der Schaaf would carry out the technical construction of the nitriding furnace and the gas flows (of NH_3 and H_2), which had to be done "from scratch". He and others in the group of Korevaar were apparently not used to embarking on a project with energy and perseverance. In my eyes his progress was much too slow and I became more and more impatient. Eventually I exploded and addressed all my anger to him. This could have been the end of any cooperation of the two of us. However, the reverse occurred: Hans now concentrated on his assignment and together we were able to arrive at an apparatus satisfying the requirements. The success we had with our subsequent investigations, also leading to the presence of his name on a number of our publications, much stimulated him (and us). We became to respect and like each other and no serious problem ever occurred again between us. To my great sadness, Hans died from cancer a few years after I had left the group.

---

[7] The "chemical potential of nitrogen" is a notion from thermodynamics and provides a quantitative measure for the "nitriding power" of the (gas) atmosphere applied.

Microscopy

A few years later the Dean of the faculty, apparently aware of my success with "reanimating" more or less under-productive coworkers of the Laboratory, asked if I would be willing to assume the supervision of the light optical microscopist of the Laboratory, Pieter Colijn. Light microscopy is an important initially microstructure characterizing technique in materials science. I knew Pieter already from my Ph.D. time and remembered all too well how difficult it was to get, in an enduring way, sustained help from him. Yet it was evident that he was a very good microscopist. I decided to talk with Pieter first and see if he would accept me as his boss. He accepted, as he himself saw a chance for him.

We started with applying all known light optical microscopical techniques and learning the theory behind them; Pieter only partly mastered the latter aspect, although he always gave others the impression to know it all. Together we spent many hours, instructive to both of us, in the darkened laboratory where the microscopes stood. While viewing in the dark through the eye-piece of the microscope and making prints of the micrographs and interferograms, we spoke also about much more than microscopy alone; fresh in my mind are conversations about music, where our tastes had much in common.

It is difficult to express in words the atmosphere, experienced by a microscopist, of being in the dark, often alone, and looking at the exposed, tiny microstructural, even atomic details revealed by a transmission electron microscope (TEM) or, at lesser magnification, by a light microscope (LM). The sensation that I got was not only that of beauty: Seeing, with the TEM during my Ph.D. time, the dislocation[8] network at the interface of a copper/nickel bicrystal and observing, upon annealing the specimen in that TEM, the dislocations move, rearrange and becoming annihilated in real time, filled me with the awe connected with the thought that I was seeing, "live", the inner workings of nature. We recorded, while the anneal ran, these structure changes on video tape that not only contained a track

---

[8] A "dislocation" is an error in the, ideally completely regular, arrangement of the atoms in a crystal. The mistake is "one-dimensional" and therefore one speaks of a "line defect". In the TEM images discussed here, with the viewing direction perpendicular to the interface of the copper/nickel bicrystal, each dislocation is indeed visible as a line. In the interface of the bicrystal, where both crystals have the same orientation in space but possess slightly different distances between the atoms, the dislocation network at the interface compensates for the differences between the interatomic distances at the copper side and the interatomic distances at the nickel side.

for image recording but also one for capturing our words spoken in a microphone, which assisted us in the later interpretation. An enthusiastic student involved in my project regularly stayed till deep in the night while performing these TEM experiments. I was surprised to hear on the recording next day not only his words informing about the instantaneous technical details, but also his humming and softly singing: he was having happy moments. I well understood how he must have felt.

Whereas the images obtained in a TEM are of the black/white type, microstructure analysis with LM has no such limitation. Especially after dedicated etching of the looked at specimen cross sections, beautifully coloured images can be observed. Especially polarized light microscopy, phase contrast microscopy and differential interference microscopy all can give rise to such wonderful experiences. Dark field microscopy can produce images where small grooves (etched grain boundaries) and small (height) interruptions, on for the rest smooth surfaces (on the cross section of the specimen investigated), all appear "light" against a deeply dark background, and can thereby induce the feeling of perceiving illumination decorations as observed after sunset in towns during Christmas time.

Light microscopy is a very powerful method to start any microstructural investigation. It is also nowadays underestimated by many. We demonstrated the various techniques with specific specimens and also applied it to our nitrided specimens.[9] Here Pieter Colijn became a real master as he revealed details which had not been seen before. We published a paper, with Pieter as first author, which influenced others in the field of nitriding. Pieter and I could very well go along with each other and I remember this experience with pleasure and gratitude.

The results of our research on nitriding were also presented on the regular meetings of the DAF, Philips, TUD and TUE working group indicated above. Obviously strong interest especially existed for our data obtained on the development of residual stress-depth profiles upon nitriding, in view of the thought relationship with the fatigue resistance (see the above intermezzo "*Nitriding*"). After having overcome my reluctance, expressed above, and because it was desired within this working group, my first Ph.D. student on nitriding, Herman Rozendaal, had, next to his fundamental research, also

---

[9] It is worthwhile mentioning that we did most observations with a Neophot 2 microscope. This microscope originated from the Deutsche Demokratische Republik (DDR) and was one rare example of a product of that country that could successfully compete with manufacturers from the non-communist world. Thus, at the time, Neophot microscopes could be found in many university and industrial laboratories at our side of the "Iron Curtain".

performed measurements on nitrided steels. The results yet appeared interesting as the steels chosen were very different (carbon steel vs. distinctly alloyed steel) and so were the results, and precisely therefore these results lent themselves to some useful conclusions.

I learned that, to reach a more engineering oriented audience, combining fundamental research on model systems with experiences from practical systems draws considerable interest. I soon felt well and fully integrated in the working group and developed a very friendly relationship with a.o. Harry van Lipzig and Jacques Hertogs (both DAF) and Theo van Heyst and Ad van der Grijn (both Philips).

German Versus, the Lingua Franca of Science, English

Although the *lingua franca* of science and engineering is English, it surprised me in these years to find out that there were two areas of my research where many important publications still appeared in the German language. Consequently, this work remained largely unnoticed in the non-German speaking world.

Indeed, nitriding was significantly developed technologically in Germany and results from those and previous days were often published in the solely German-language journal "*Härterei-technische Mitteilungen*" (*HTM*). The journal now is named "*Journal of Heat Treatment and Materials*", where, since a few years, each article appears both in English and in German. Many years after the events recounted here, in 2000, I would become Editor of the journal and remain that for 20 years. In the past many of the significant papers published in this journal were presented first at the yearly held conference "*Härterei-Kolloquium*" in Germany, an event with then of the order 800 participants coming not only from countries with Germanic languages, but also from, for example, the USA and Japan.

After I had been active on nitriding for a few years and had published (in English) some work, the DAF etc. working group suggested that I should present my work on the forthcoming Härterei-Kolloquium in October 1980. The meeting was held in the Kurhaus of Wiesbaden. It would be my first major lecture in the German language and I was pretty nervous (even now, after having lived for many years in Germany, my English is still somewhat better than my German…). My nerves were even more strained after finding out not only that the audience indeed was enormous, but also that there was a large green light, very visible to the entire public, mounted on the pulpit, that turned orange five minutes before the end of the allotted lecture time and that finally turned red and *started to rotate dramatically* if all lecture time had been consumed, causing hilarious moments in the audience, as I noticed.

That embarrassment I wanted to avoid in any case. Happily, I succeeded and could enjoy positive responses. Especially the Dutch-Belgian(Flemish) community, including the DAF etc. colleagues, that each year attended the meeting, was visibly "proud" that at last somebody from the Low Countries had been lecturing at the meeting. This group of people traditionally then met in the evening/night in a winery downtown Wiesbaden. That night I joined them and there is a photograph of that occasion showing that I must have been pretty relieved after the just described exertion.

Five years later, in 1985, the same Dutch-Belgian community founded the Dutch-Belgian Vereniging voor Warmtebehandelingstechniek (VWT; Dutch-Belgian "Society for Heat Treatment"). I was asked to be their President and I remained as such for 10 years. On the occasion of my retirement from this position I was appointed "Honorary President". Coming from pure science and largely staying there also during the here discussed period of time, this contact with the more engineering side of materials science has been a most rewarding facet of my life. There is no doubt in my mind that scientists should now and then cross the "boundary" to practice/engineering, because they then can contribute essentially. Moreover, I have met in this circle of engineers/scientists more heartedly felt friendliness than in the circle of my colleagues in the more or less purely scientific, later half of my professional life.

The second area, where I have been active and where papers published in the German language dominated the scene, was the (X-ray diffraction) analysis of (residual) stress. In my younger years, which are discussed here, two German Professors, Viktor Hauk and Eckard Macherauch, were the established "popes". Their work and that of their co-workers was regularly published in German, indeed also in *Härterei-technische Mitteilungen* (see above). Again, these to large extent purely scientific, methodological results remained for a long time largely unknown outside the Germanic countries. This led Jerry (J.B.) Cohen at the Northwestern University (Evanston Illinois, USA) to invite H. Dölle, who had just finished his Ph.D. project under supervision of Hauk, to America for a postdoctoral stay, with the effect to introduce the "German" results into the English speaking world. More or less as a result, Cev (I.C.) Noyan, co-worker of Cohen, together with Cohen, later published a book (*"Residual Stress; Measurement by Diffraction and Interpretation"*, Springer, 1987). Thereby, the foregoing, fundamental German literature for the first time was widely appreciated in a non-German publication by non-German authors. Later, also in Germany more and more publications in this field appeared in the English language (Hauk, as editor and with only German co-authors, published an extensive book on this topic in English (*"Structural and Residual Stress Analysis by Nondestructive Methods; Evaluation—Application—Assessment", Elsevier, 1997*)). All my papers originating from the work on this topic in Stuttgart have been published in English (in journals as the *Journal of Applied Crystallography* and the *Journal of Applied Physics*). Yet, a few German groups still present such work in German as well… Germany until the present day still has a leading position in this field.

Unusual Lattice Parameters

My years in the group of Korevaar did not focus on "nitriding" alone. The initial and later stages of precipitation of second phases in aluminium (Al) alloys were investigated. I had chosen two alloys exhibiting very different precipitation behaviours: aluminium–silicon (Al–Si; stands for the class of aluminium casting alloys) and aluminium-magnesium (Al–Mg; stands for the class of aluminium alloys utilized after precipitation hardening). Whereas in Al–Si alloys, supersaturated with dissolved Si, more or less direct precipitation of the end, equilibrium phase (pure Si) occurs, in Al–Mg alloys, supersaturated with dissolved Mg, a precipitation sequence is observed: the precipitation process is initiated by the development of local, disordered enrichments of Mg while the Mg atoms remain on sites of the Al crystal structure, which enrichments are called "clusters", and only upon continued aging/annealing intermediate precipitates and the final equilibrium precipitate can occur. The reason to briefly mention this work here is the following: in a research project it can be very useful to apply a strategy where two or more systems, showing contrasting, even opposite effects upon subjected to similar treatments, are investigated in parallel, as thereby more insight in the essence, i.e. a more comprehensive, general understanding, of the phenomenon investigated, may be obtained. I deliberately applied this approach a few times in my career as a research leader and met with success (also see the above text about research on nitriding applying two microstructurally and compositionally very different steels).

One piece of this research on aluminium alloys had, especially in time and scope, wide ranging consequences. Upon completed precipitation of silicon in an aluminium–silicon alloy, the specimen consisted of a practically pure aluminium matrix (in equilibrium aluminium has negligible solubility for silicon) containing fine precipitates of pure silicon (silicon has negligible solubility for aluminium). In the X-ray diffraction pattern of such a specimen one would expect reflections from aluminum and of silicon. Indeed this was observed. However, we found out that the lattice parameter (a measure for the distances between the atoms) of the aluminium, as determined from the reflections in the diffraction pattern, was clearly larger than that known for pure aluminium. How could that be? First it was ruled out that contaminations played a role.

By sheer chance, at that time I was reading in/studying parts of a well-known book for physical materials scientists: *"The Theory of Transformations in Metals and Alloys"*, written by J.W. Christian (2nd edition, Pergamon Press, Oxford, 1975). My copy of this book was a gift, from the Okkerse group on the occasion of my leave for Korevaar's group, which turned out to be a present highly valued by me.

In this book ample attention is paid to the theory of Eshelby describing the elastic strain/dilatation effects caused by "point imperfections" in a crystalline matrix. With "point imperfection" is meant, for example, the presence of a foreign atom of size different from that of the atoms of the parent crystal. This size difference implies a certain misfit at the location of the foreign atom: the foreign atom does not fit well into the space available at the atomic site of the parent crystal now occupied by the foreign atom and thus elastic deformation is induced. The theory allowed for the calculation of the effect of the presence of such foreign, misfitting atoms on the average lattice parameter (i.e. a measure for the distances between the atoms) of the parent crystal. Eshelby had hoped that this theory would provide good predictions of the change of the lattice parameter observed upon dissolving a foreign element B in a piece of crystalline, initially pure element A. However, no agreement of prediction and measurement could generally be obtained. The reason is that on atomic scale, i.e. on the scale of a "point imperfection", electronic interaction can dominate elastic straining effects, thereby invalidating application of elastic deformation theory. However, this objection no longer holds if much larger inclusions (much larger than a single atom) are considered. I thought that the yet tiny, but much larger than a single atom, silicon precipitate in the aluminium matrix could be such an object compatible with now allowable application of the Eshelby theory.

Following the described approach led to success: the difference of the measured (larger) lattice parameter of the aluminium matrix, containing the misfitting, tiny silicon precipitates, with the (smaller) lattice parameter of pure aluminium very well agreed with the result calculated adopting the theory indicated above. To our knowledge this was the first observation and explanation of dilatation (here crystal-structure expansion), expressed as change of the average lattice parameter, of a matrix containing misfitting inclusions. The result was published in 1981. Thereafter, time and again during my career, I unintentionally met scientific problems that asked for application of this theory:

A few years later, I used the same approach to explain quantitatively the "too large" lattice parameter of pure iron containing misfitting iron nitrides. Still later, in Delft and thereafter in Stuttgart, the theory provided background to understand the uptake of dissolved nitrogen in iron beyond the equilibrium solubility for pure iron, as a consequence of the presence of misfitting alloying element nitrides, e.g. chromium nitrides, in the iron matrix.

Nearing the end of my career, in Stuttgart, I decided to return in a general way to this topic of lattice-parameter change of a matrix, due to the presence of misfitting inclusions as second phase particles. Together with two Ph.D. students, Maryam Akhlaghi and Tobias Steiner, and a co-worker, Sai Ramudu Meka, we gathered all cases known to us where data on lattice-parameter changes of systems with misfitting inclusions were available and introduced the distinction of systems where coherent (dependent) diffraction of the inclusions with the matrix occurs and cases where the inclusions and the matrix diffract independently, i.e. incoherently. I had noticed in the many years passed since 1981 that the generally distinct influence of misfitting inclusions, as precipitates, on especially lattice-parameter values often was not realized, which could lead and had led to strikingly erroneous interpretations. This was a motivation for the final overview paper published (*Journal of Applied Crystallography, 49 (2016), 69–77*).

After not more than a couple of years in the office on the first floor of the wing with the Halls, I moved to an office on the ground floor of the main wing, close to the entrance of the Laboratory. I now was very close to our laboratory with the nitriding furnace and other, dilatometric, thermogravimetric and calorimetric equipment, which was situated directly at the right-hand side upon entering the Laboratory and thus more or less opposite to my office, across the corridor. That enhanced the directness of the interaction with a.o. Hans van der Schaaf and the students working there.

Professor Meijering

One day, probably in 1981, the Dean of the faculty, Professor Kievits, asked me to come to his office. He asked me if I would agree with sharing my office with Professor Meijering, who had retired and now needed a writing desk, etc. and some space as an emeritus. This question surprised me enormously. I would have expected that Meijering as an emeritus would be allowed a small office of his own (Professor Burgers, also retired, but also still regularly at the Laboratory, had one). Kievits had two arguments: Firstly, there was no unoccupied office available. Secondly, it would be very worthwhile for me to be able to have discussions with Meijering. The second argument was certainly a highly unusual one; I did not know of a retired Professor sharing an office with a young scientist only a few years after his Ph.D. time.

Professor Meijering (another Professor appointed in Delft coming from the Philips Research Laboratories; see Chap. 5) had contributed fundamentally to the understanding of binary phase diagrams and had been instrumental in the development of the calculation of ternary phase diagrams from the component binary diagrams.[10] That work was highly esteemed worldwide. Moreover he had published important, widely cited research on the internal oxidation of metals. This work would become of importance to me in my work on nitriding (but Kievits could not know that).[11] During their active years Burgers and Meijering were certainly the most reputed scientists, of worldwide standing, within the Laboratory of Metallurgy.

I knew Meijering, of course, but had had no other contact with him than on the occasion of the defense of my Ph.D. thesis, where he was a member of the examination committee. I also had seen him on several occasions in the laboratory with the light optical microscopes (the "kingdom" of Pieter Colijn; see above). He was certainly the only Professor in the Laboratory I had ever seen sitting a few times behind the microscope and intensely studying the microstructure of a specimen, which involvement I judged very positively. Hence, Fig. 8.5, showing Meijering behind such a(n old) microscope, presents an image of a reality that was not only posed.

I accepted the proposal by Kievits: I would certainly enjoy and could learn from discussions with Meijering, and, moreover, Meijering was a sympathetic and humorous man. Unfortunately, Meijering did not come often to our joint office and seemed a bit shy, which formed a barrier to initiate a discussion. Yet, a few things, he told/learned me, stayed with me all my life:

---

[10] A phase diagram shows the fields of stability for phases as a function of (so-called "intensive" = independent of the size of the system) state variables, as pressure, temperature and composition. Thus, for a binary (=two-component, say A and B) system at constant pressure (usually 1 atm) the phase diagram is presented as a two-dimensional drawing with the temperature along the vertical axis (=ordinate) and the, in this case only, composition variable (say, the amount of B in the alloy A-B) along the horizontal axis (=abscissa).

[11] "Internal oxidation" of a metal alloy, say the alloy A-B, with B as an element of strong affinity for oxygen and A as an element of relatively (very) weak affinity for oxygen, can lead to the precipitation of small particles of the oxide BO_n in the matrix of A upon inward diffusion of oxygen (O). By substitution of oxygen (O) by nitrogen (N) in the preceding sentence the phenomenon of "internal nitriding", sometimes called "internal nitridation", leading to the precipitation of tiny particles of the nitride BN_n in the matrix of A, is similarly introduced.

Fig. 8.5 Prof. Jan L. Meijering "in action" at a classical light optical microscope in the Laboratory of Metallurgy (photograph from the repository *"Beeldbank TU-Delft"*, *1962 (Delft University of Technology)*

It often happens during the drafting of a paper by more than one author, that a heated discussion arises among the authors about a relatively small text part, for example dealing with (interpretation of) a certain result from an experiment or a theoretical argument. In that case, Meijering said, the science discussed in that text part is very often unclear, not well/not deep enough understood. Instead of to continue the debate and fight it out and keep text in the paper that is likely criticized easily by others, it is simply better to delete such text, to avoid damaging of the impact of the more important, larger part of the paper, and perhaps also to exclude bad mood among the authors.… I have followed this advice on many occasions and repeatedly taught my students in the same way.

Meijering was very clever in handling mathematics and was a master in using "handwaving" physical arguments. At one time, I do not remember why, grain-boundary energies[12] were discussed by us. Their values are experimentally extremely difficult to impossible to obtain in a direct way. Meijering had published a paper where, almost as a side remark and on the basis of a certain, one may say blunt justification, he proposed to estimate a grain-boundary energy as the energy required to remove $\frac{1}{4}$ of the atoms from a crystal-structure plane; thus $\frac{1}{4}$ empty crystal-structure sites result (such empty sites are called "vacancies").[13] This rule of thumb I applied a few times years later and, for me not unexpectedly, found out it is an efficacious one.

In retrospection it must be said that my scientific exchanges with Meijering were of limited frequency, but I gratefully acknowledge that their impact was lasting.

---

[12] A piece of e.g. a metal is usually constituted of many crystals tightly connected with each other so that a massive piece of metal results. These usually small crystals (to be observed under a light optical microscope) are called "grains". The boundary between two such crystals/grains is called "grain boundary". The atoms at the grain boundary tend to comply, at the same time, with the prescriptions for the locations of their sites on the crystal structures of *both* adjacent crystals, which generally of course is impossible. This can lead to a certain disorder of the atomic arrangement at the grain boundary. As a consequence the atoms at the grain boundary, experiencing a state of less desired bonding, have a higher energy than in the bulk of either one of both adjacent crystals. This enhancement of energy is expressed by the so-called "grain-boundary energy" (also see "*Interface and Surface Energetics; a Materials Science Approach; wrong "Diagnostics""* in Chap. 13).

[13] To be accurate, the assessment of the grain-boundary energy by the rule of thumb given pertains to "incoherent" grain boundaries, where the adjective "incoherent" indicates that, upon passing from one crystal to the adjacent one, there is a discontinuous and not a continuous, gradual, smooth change of the atomic arrangement at the grain boundary.

"Particularities" of the Contacts with Polish and DDR Scientists

During the years in Korevaar's group I had various intensive contacts and cooperation with scientists from behind the Iron Curtain (the Fall of the Iron Curtain occurred later: in 1989). One of these was Ignacy (called "Jacek") Wierszyllowski from the Technical University of Poznan in Poland. He spent a year with me in Delft and it was he who brought me to the study of iron–nitrogen martensite, which started a whole new line of research that I later followed after my return as Professor to the former "Burgers group". Jacek wanted me to visit him and his laboratory in Poznan. So it happened: in October I went by train via divided Berlin to Poznan (in German: Posen). For me this became a bit of adventure, as the circumstances in Poznan were totally different from what I was used to in Western Europe.

It started with the notorious East-German police officers, "VoPos" (with "VoPo" is meant "Volkspolizei" = People's Police), in keeping up the train in Berlin for more than six hours, without any apparent reason, causing a strongly delayed arrival in Poznan in the middle of the night. Jacek was there waiting for me on the platform and brought me to the guesthouse of the university. It struck me that it was extremely hot in the room and that it appeared impossible to reduce the level of heating. The heating system operated at full power, day and night, although outdoors it was about 20 °C. I was bewildered and asked Jacek the following day why one was heating the building to tropical temperatures, as this was an apparent waist of energy, and why I could not adjust the heating power in my room. Jacek informed me that the current "5-years plan" of the Polish government required that the heating system had to be switched on in the first week of October, indeed in fact for the whole building, and that one had to obey this requirement of the authorities, irrespective of the real weather conditions. It was my first personal experience with the idiocy of the bureaucracy of countries behind the Iron Curtain.

A strongly penetrating, sharp odor of lysol entered the nose wherever I went in the guesthouse or in the laboratories of the university. Even walking the streets of Poznan, a city with buildings and houses along the streets reminiscent of the long time the city was Prussian ("Poznan" in German is "Posen"), one could discern this smell. Lysol was clearly the favoured cleaning agent, sanitizer and detergent. In these days, nowhere in Poland you could escape its smell.

I was aware that the welfare state as I knew it could not be found in Poland. Yet, it shocked me that at the lavatory in the laboratory of Jacek, after having relieved him/herself, one was expected to clean him/herself with newspaper pages cut into strips. On the occasion of my next visit I brought with me a few rolls of toilet paper…. and thereby did not only myself a favour.

The laboratory of Jacek covered a couple of rooms of an old, decrepit villa from Prussian times. The scientists in Poland in these years experienced great difficulties to acquire experimental equipment at top level from western firms. And in the particular case that they did, servicing by technicians from the producer was beyond the financial means of my Polish colleagues. Jacek, for example, possessed thermogravimetric equipment made by Setaram, a high quality instrument from a French firm, but he was unable to use it at the desired level as the instrument was not in appropriate condition. I suspected that this general deficiency may have led to a lack of accuracy and precision in the experimental work in general in the Polish laboratory. I had difficulties in transferring to my Polish colleagues the attitude in the Laboratory in Delft where the instruments used were first brought to the level where the accuracy inherent to the instruments was achieved and reproduction was checked and guaranteed. However, our combined efforts eventually led to a joint paper in 1983 on iron–nitrogen martensite.

One time in Poland Jacek and I planned to attend a conference at the Baltic Sea coast in a town called Kołobzreg (perhaps better known in German as Kolberg). We would travel with Jacek's car. However, on the morning of our departure from Poznan Jacek "discovered" he had not enough petrol. This substance was hard to get these days in Poland. Then I observed how important "networks" in East-European countries behind the Iron Curtain were: in no time Jacek had called some friends and relatives and was brought some canisters petrol. I would witness this sort of mutual assistance and direct help also on other occasions in this country and the DDR. The social cohesion in these countries was distinctly larger than that in West-Europe. Admittedly, it had developed out of necessity, but it was a virtue which is nowadays regrettably in decline, or has even become out of sight entirely, in our western society being focused on individual "self-optimization".

Having thus fueled the car, off we went. On our way heading north we passed through a country side characterized by green hills and small villages in the valleys. A thick layer of yellowish-greenish fog was stretched as a blanket over the houses in the villages. Jacek told me this was caused by using peat as a fuel for domestic heating and cooking. I had never witnessed myself such use of peat and thus the dense fog it caused. The last time peat was utilized in The Netherlands for domestic purposes was already very many decades ago and, indeed, it was also a characteristic of poverty, as the peat diggers, living in the so-called "peat settlements" (in Dutch: "Veenkoloniën", a notorious term) were very poor people.

The weather was bad, so late in the year, and the Baltic Sea could not impress me; I was used to the much larger rollers of the North Sea. But the conference was very nice.

It was almost impossible for scientists from countries behind the Iron Curtain to participate in conferences not in their own country: supposing they had the funds for attending conferences in the "free world", they would not get permission for leaving the country, out for fear of their defection..... If you met such a scientist at a conference in the USA or Japan or in Western Europe yet, you could take it for granted that he/she somehow was connected with the ruling class/party of their country. But it was less difficult, although still not easy, for an East-European scientist to attend a conference in another country of the communist bloc. So, in Kołobzreg, I met a few scientists from the DDR (=Deutsche Demokratische Republik; i.e. so-called "East-Germany").

Whereas the Polish scientists were rather completely "open" to me, this did not hold for the DDR scientists. Only after some time I understood the reason for this: I got acquainted with a scientist from the DDR and after a couple of longer scientific discussions, at a few days, we also touched upon other themes. An atmosphere of trust developed. It was thus revealed to me that in the delegation of DDR scientists to conferences there was always a non-scientist employed by or with close contact to the secret police of the DDR (i.e. the "Stasi", which is a short notation for "state security", formally the "Ministerium für Staatssicherheit" of the DDR), whose assignment was to control and to report on (!) the genuine scientists of the delegation. So it was also the case at the Kołobzreg meeting. Hearing this, a feeling of the always present threat, as must have been experienced by the DDR scientists as hanging over them during their daily life, gradually was becoming more vividly clear to me. These times are now, after 1989, over and a lot of information about this past is available. We may yet still be unable to understand and *sense* how it must have felt to be a scientist under such circumstances.

Shortly thereafter I got an invitation to attend as key lecturer a conference in the DDR at Karl-Marx-Stadt, nowadays again called Chemnitz, as before DDR times. Tom Bell, a Professor at the University of Birmingham, who I did know well (see Sect. 9.3), would also lecture there. I accepted. Now I needed a visa, which was more complicated to get than I thought. I had to come personally to the DDR embassy in The Hague and had to inform them about a lot of personal details and moreover specifically not only when, but also how (by train) and especially *where* I would cross the border between the Bundesrepublik and the DDR.

Sitting in the train standing at the BDR/DDR border, VoPos went through the train, checking everybody personally and carefully. When it was my turn, my name was checked against a list the officer had and indeed I was on the list. I was more than flabbergasted that the bureaucracy of the DDR managed to get my name on the list of that day for the VoPo in precisely that train I was travelling in, at the location at the border that I had to inform the embassy in The Hague about. It gave me the correct impression of how life was top-down controlled in the DDR for its citizens.

In Karl-Marx-Stadt I met Werner Schröter, who was my actual host. He picked me up from the hotel, likely the best hotel in town, but where, perhaps typifying the nearly defunct state of the DDR already then, the elevator did not work during my whole stay, so that I had to use the stairs up to the fourth floor where my room was. Leaving the hotel I saw Werner at the other side of the street and, as there was no traffic on the street at that moment, I immediately crossed the street to meet him. Only to observe great anxiety on his face, that he declared to me: I had crossed the street without using the zebra and moreover ignoring the red light for the pedestrians. He said the police was very, very strict and one would be in real trouble if caught. DDR citizens did not dare to do what I had just done. This enhanced my feeling of unease, which I had already. Poland was almost "liberal" as compared to the DDR.

At the conference Tom Bell lost his passport and had, imaginable in view of the above experiences, great trouble to get papers allowing his return flight to the UK. He said he would never again pay a visit to the DDR. If he kept that word, I don't know; after all Tom was an excessive traveler.

With Werner I had some interesting discussions about his and our results on nitriding. His work was largely published both in German and in DDR journals, which was a barrier for dissemination of his results and ideas on the international platform. He was the one who had introduced the word "molnite" ("Molnit" in German) for the nitrogen *gas* (N_2) phase that precipitated in nitrogen supersaturated iron–nitrogen phases, recognizing the parallel with the precipitation of *solid* graphite ("Graphit" in German) in with carbon supersaturated iron.[14]

---

[14] The iron-carbon (Fe–C) and iron–nitrogen (Fe–N) systems are of great technological importance (steels!). They show a lot of parallel phenomena, but also a number of significant and, for practical applications, important differences. Thus their phase diagrams (see footnote 10) exhibit to a certain extent a similar morphology. However, the stable carbon phase at room temperature and at normal pressure(1 atm) is the *solid* graphite, whereas the stable nitrogen phase at these conditions is the *gas* nitrogen. The nitrogen gas phase (N_2) can precipitate in solid iron–nitrogen phases, as with dissolved nitrogen supersaturated iron and iron nitrides, and thereby these phases become porous. This porosity is generally detrimental for the mechanical properties. Precipitation of a gas phase in a solid matrix is a rather unusual phenomenon. Thus it can be understood that the pores, which can coalesce to

I noticed that in Poland it was possible to acquire, for a small amount of money, long play records, LPs, with recordings of the highest musical quality (the first compact discs, CDs, were about to appear on the market but not generally available; they became popular in the second half of the eighties). Thus I took with me, back to Delft, an LP with Schubert songs performed by Theo Adams (a bass-baritone from the DDR) and Jörg Demus (a pianist from Austria) and an LP with a rare live recording of Swjatoslaw Richter (a Russian pianist). I still have these records and play them if I can find the energy to start up my old grammophone player… All these three musicians are no longer alive. Hearing Theo Adam sing "An die Musik" does not cease to strongly move me.

The Grand Piano

I had let move the second hand, upright piano, with which I had learned piano playing (see at the end of Chaps. 2 and 3), from the house of my parents in Haarlem to my apartment in Delft. My brother had not expressed great interest in piano playing. The piano lessons were a burden for him. Especially the "quatre mains" pieces, which we had to play as boys, often led to bickering between the two of us…. So it was allright I got the piano.

The piano was not bad, but also not very good. Especially its touch was not light enough in my eyes. It was my dream already for many years to possess and play on a grand piano. My grandmother (from mother's side) had a girlfriend, of course also already an old lady, in Apeldoorn, with whom she regularly had contact. I suppose my grandmother was familiar with her from the time she lived in Zutphen, not very far from Apeldoorn. This lady was a family member of and close to the owners (till 1978) and the managing directors of the paper factory "Van Houtum en Palm" with factories in Apeldoorn and Ugchelen. This firm had a certain elevated status as it produced the high quality paper for the banknotes of the Dutch money. I am not sure anymore if the girlfriend of my grandmother was a "van Houtum" or a "Palm"; I will call her here "lady Palm". She possessed a grand piano, a Steinway, as my grandmother knew, but didn't play the instrument any more. It must have been presumably between 1976 and 1978 that my grandmother, knowing my wish, said to her that she had a grandson who could play reasonably well

---

channels in contact with the outer atmosphere of the specimen, have long been subject of debate regarding their cause. Nowadays the origin of the pores, as described above, is generally accepted (also see footnote 2 and the associated remark in the main text in Sect. 12.1).

the piano and had the strong desire to acquire a grand piano. After consultation of lady Palm with her children, who then agreed with "giving up" (from a possible heritage) the grand piano, I was asked to come to Apeldoorn with my grandmother and to play the piano, as participating in an audition, so to speak.

From the moment I was seated behind the keyboard, somewhat nervous, and started to play I was thrilled. Everybody who for the first time plays on a real and high quality grand, being used to an upright piano, will have the same experience as I had: the timbre and, indeed, light touch give an overwhelming, lasting impression. Apparently my playing convinced lady Palm that I could be a deservedly player of the instrument. We agreed a price, which was modest, but a large sum for me.

That is the way I got my "Steinway", thanks to the mediation and gentle pressure of my grandmother. The grand was built in Hamburg in 1950 and thus is as old as I am. I still have this instrument and it is still in good condition, albeit needing small repairs and optimizations now and then, apart from regular tuning, of course.

The End of my Stay in Korevaar's Group

After Okkerse had left the Laboratory to become the (deputy- then) Director-General at the Ministry of Education and Science, the Burgers Chair remained unoccupied considerable time until late in 1979 the appointment of Sijbrand (Sieb) Radelaar was realized. Radelaar was a physical metallurgist by education and had already gathered professional experience on a couple of academic locations.

Radelaar's background certainly fitted well to the former Burgers/Okkerse group. Whereas Radelaar had built up a group of his own at his previous university affiliations (the University of Groningen and the University of Utrecht), the group in the Laboratory in Delft already had a very good international reputation and in view of this recognition did not obviously need drastic restructuring. Imposing a pronounced strategic change was not easy. Here he was faced with the same situation that may have been an additional reason for Okkerse to leave the Laboratory, as I wrote in Chap. 5. Yet, in the relatively short period of time that Radelaar was there, he put a few distinct accents in research on layered structures, especially silicon-based ones (for example of technological importance in view of their application in integrated circuits) and on amorphous materials, especially metallic glasses, which was research he already had performed at the University of Utrecht (his immediately preceding affiliation). The latter work led to a rare example of direct cooperation with the group led by van den Beukel in the Laboratory.

Radelaar was the one who once told me, that my time in the group of Korevaar should be considered by me as a blissful time. In next steps of my career, to be much more constrained by managerial responsibilities, he said, such happy times in a researcher's life would not return to such extent. I was rather soon about to find out that he was all too right.

Radelaar had a dynamic, even restless personality, which did not allow him to focus a longer time on one scientific topic or, in a broader sense, on a scientific field, or even on science at all. In the course of his career he accepted many managerial assignments and was successful there. After only a few years, he left the Laboratory in 1982 to become Director of the Center for Submicron Technology at the Delft University of Technology. It was not his last affiliation.

After some years in Korevaar's group I started thinking about my future in a more direct way than before. I knew that Professor positions were difficult to acquire, already only given the rarity of a vacancy.[15] Moreover, mere chance and luck also were crucial: "to be the right man or woman at the

---

[15] In Germany the situation is different: after having written a "habilitation" (something as an extended Ph.D. dissertation) and having served a few years as "Privatdozent" (="private lecturer"/"adjunct professor"), implying to have given a lecture course, one more or less automatically is granted the right to carry the title "Professor", albeit with the addition "apl." meaning "ausserplanmässig" (="unscheduled", here: not part of the Professors in the faculty holding a Chair). Therefore the number of Professors in Germany is relatively very large. (In Austria the number of Professors is (also) overwhelmingly large. I once met an Austrian commenting to me: *"In Österreich ist jeder Schauspieler Professor"* (= *"In Austria every actor is professor"*)). The decay of the Professor title is emphasized by former polytechnics and the like (i.e. schools offering professional, not academic educations) being elevated to the category university involving that their lecturers call themselves Professor. An ultimate sign of "title illness" then occurs if these nonacademic schools grant a person the title Honorary Professor. Thus I observed, to my amazement, that the Lord Mayor of a city in Germany suddenly had begun to decorate his name with "Prof.". I soon found out that no scientific achievement could be indicated allowing such honour; in this case it was a honorary title that was granted by such a school as meant above, moreover a private one, of which I had never heard of before (but that is my fault).

As a consequence, a distinct hierarchical system regarding professor positions has developed in Germany. A highly amusing booklet, actually a caricature but not too far from reality, presenting the workings of this stratified organization of scientific, also Professor, ranks, is: *S. Bär, Forschen auf Deutsch, der Machivelli für Forscher – und solche die es noch werden wollen* (=*Research, the German way, the Machiavelli of scientists – and of those who desire to become one), Verlag Harri Deutsch, Frankfurt am Main, 1992.* I got a copy of this booklet from a respected, colleague scientist at Degussa in Hanau, while I was still active in Delft but after it had become known I would go to Stuttgart, and a second copy of the same booklet I got from the Works Council of the Institute on the occasion of my start in Stuttgart. Evidently, one thought I needed some "education" before "diving" into the German scientific world.... In the Dutch system, more or less equal to the Anglo-Saxon system, the Professor title was more a rarity and, apart from the few Honorary Professors, reserved to Chair Holders at moreover genuine universities.

right moment". So mentally, I prepared myself to stay for a longer time as "wetenschappelijk hoofdmedewerker" (I was recently promoted to that rank, something like "associate professor"), but was not prepared to stay in my current job for "eternity" and would have even accepted a job in industry (where, in the meantime, I was not unknown) before I would have been "too old".

As far as I know (the deliberations of the nomination committee remained a secret to me), I was not immediately considered as the first candidate for Burgers Chair after Radelaar had left. My disadvantage, of course, was that I would be something like an in-house appointment, something that was rare and undesired at a university of technology, where preceding industrial experience and at least experience at another university or research institution was considered as a "conditio sine qua non". What spoke in favour of me were my research record, leadership capability, young age (at the same time that could also be considered a disadvantage) and a research profile that perfectly fitted to the Chair. Although I desired the position, I did not consider that, in the end, I would be asked to be Professor on Burgers Chair. Yet, that is what happened.

So in 1985, at the age of 35, as their Professor I returned to the group where I had concluded my Ph.D. in 1978.

9

"Young" Versus "Old"

Abstract A "young", rather unknown scientist, with a finding/idea overthrowing established knowledge, more than rarely meets enormous resistance by the "old" scientists constituting the "elite" in the field of science concerned. This chapter recounts the history of three such "fights": the discovery of "quasicrystals", the "Kirkendall effect" and a debate on the occurrence of "clusters" or "precipitates".

Some ideas only die with their proponents

Time and again the advent, in a certain field of science, of a new idea or a new finding, distinctly deviating from what is considered by the instantaneous "elite" in that field of science as the "truth" or "reality" of nature, meets enormous resistance. From the proponents of the new insight this requires great tenacity to persevere and eventually to overcome the resistance. As long as such discussions and debates solely concern exchange of arguments and are finished by agreed upon, accepted conclusion this may typify how science "runs". However, in many cases human character traits may interfere destructively with such course of scientific debate. As I wrote in Chap. 6: "*those who are respected and have a high position and "live within" a certain paradigm, possibly based on own former work, are unwilling to become taught otherwise, as this is felt as humiliation associated with loss of status*". This then can be enhanced by the superiority felt by the well-known, perhaps even famous, high ranking senior (the "old" one) against the unknown, low

© The Author(s), under exclusive license to Springer Nature Switzerland AG 2022
E. J. Mittemeijer, *How Science Runs*,
https://doi.org/10.1007/978-3-030-90095-3_9

ranking junior (the "young" one).[1] These nasty aspects bring elements in the debate/discussion that not only can slow the progress of science, but, in my eyes even more harmful, they can damage the "young" scientist unjustly very severely. As a consequence high quality "young" scientists may not only leave a scientific field, but, worse, disappointedly leave science as a profession.

In the following three cases of "young" versus "old", of relevance to my own field of science and career, are discussed, which all show one or more of the features indicated above. The first one is a famous one, where the "young" scientist eventually ended up with the Nobel Prize. The second one concerns a young scientist who left science as a profession before his discovered "effect" was accepted by the peers in his field and which effect was eventually named after him. The last example concerns a personal experience where the unpleasant debate more or less ran until the death of the "old" scientist.

9.1   The Discovery of Quasicrystals; an "Impossible" Symmetry

"Normal" crystals are given by a massive, regular, periodic arrangement of identical building units, parallelepipeds. (A parallelepiped is a prism with parallelograms as faces). The simplest shape of such a building units thus is a cube. In the building unit the atoms of the substance considered occupy prescribed sites. The crystal then is obtained by departing from one such building unit and obtaining other such building units by translating/copying the unit in three directions over distances parallel and equal to the edges of the building unit, etc. Taking a cube as building unit, the operation described results in a massive arrangement of same oriented cubes sharing cube faces. The crystal thus obtained thereby exhibits *translational symmetry*. One also distinguishes *rotational symmetry*, which implies that a rotation over a certain angle about an axis through the crystal leads to the same appearance. It is now not difficult to mathematically show that translational periodicity (as characteristic for "normal" crystals) involves for these crystals that they can *not* possess five-fold (and not more than six-fold) rotational symmetry.

This was the situation as also I had learned it in the first years of the seventies as a student being trained in basics of crystallography.

---

[1] "Old" and "young" as used here do not immediately bear a relation with ages. The junior, unknown researcher opposing the establishment may already be mature and experienced. This holds for the first example discussed (Sect. 9.1). Indeed, in the second example the junior researcher is moreover rather young of age (Sect. 9.2).

It seems natural to suppose that if a liquid substance is cooled very fast that then the equilibrium state upon solidification may not be attained as the atomic mobility is too small with respect to the cooling rate; a "locked-in" metastable or unstable state results. Especially if certain, specific metal alloys are fast cooled, quenched from the liquid state, even *amorphous* metal alloys may be acquired: metallic glasses (the designation "amorphous" implies that the atoms in the metal are not regularly arranged as in a "normal" crystal as described above; instead a more or less chaotic, random arrangement of the atoms prevails). Such research on rapidly quenched metals became very popular after 1960. By pouring a stream of molten metal alloy on a fast rotating (copper) wheel the required very high cooling rates in the range of 10^4 K/s to 10^6 K/s can be achieved. This technique is called "melt spinning". Together with some colleagues around 1980 I performed such research on especially Al alloys. The high cooling rate made it impossible for the material to reach its equilibrium state upon solidification. No amorphous product was obtained, but a crystalline solid solution phase showing unusually high solubility for alloying elements which otherwise (in equilibrium) would have largely precipitated as a secondary phase.

At about the same time Dan Shechtman was interested in melt-spun metal alloys, as not only unusually high solid solubilities could be obtained, but also new, unknown intermetallic compounds could be found. During his sabbatical stay from Technion (Haifa, Israel) at the National Bureau of Standards (renamed as National Institute of Standards and Technology in 1988) in Gaithersburg, Maryland (USA), Schechtman, a very good transmission electron microscopist, (also) investigated aluminium-manganese (Al-Mn) alloys which were melt spun using the above described technique. In 1982 Shechtman, while sitting behind the transmission electron microscope (TEM), saw a diffraction pattern recorded from such an Al-Mn alloy and noticed that the diffraction pattern contained a tenfold rotational symmetry. That is incompatible with "normal" crystals; the tenfold diffraction pattern is compatible with a crystal possessing five-fold rotational symmetry.... It was and is well-known that such apparent "impossible" symmetry can occur in a crystal being the result of a specific way of multiple twinning (twinning is a special defect in the atomic arrangement in the crystal). Shechtman, by extensive investigation, excluded that that was the case here.

The observation was such extraordinary, in conflict with what at the time was considered to be the incorruptible, very basis of crystallography, that very many could not accept it: it was too unbelievable to be true. Shechtman experienced great trouble to convince. That went that far that the head of the laboratory where Shechtman worked urged him to leave; he said: "you are a disgrace to my group". Shechtman went to another group and developed the idea of an icosahedral crystal compatible with the observed "impossible" rotational symmetry. John W. Cahn at NBS had brought Shechtman to NBS for his sabbatical. Cahn was a world famous, largely theoretical materials scientist. Also he initially did not believe the result by Shechtman. Shechtman returned in 1983 to the Technion. Together with his colleague Ilan A. Blech he there prepared a paper about the unusual result. Upon submittance to the *Journal of Applied Physics* the paper was immediately rejected. Then the manuscript was submitted to *Metallurgical Transactions A*.

Shechtman visited the NBS again in summer 1984. Cahn, after having seen the manuscript by Shechtman and Blech, now was convinced of the genuinity of Shechtman's result. This resulted in a short paper authored by Shechtman, Blech, Gratias and Cahn submitted to *Physical Review Letters* (*PRL*). This paper, submitted in October 1984, was immediately published by the journal in November 1984. The original first paper, authored by only Shechtman and Blech, was published only distinctly later in June 1985 in *Metallurgical Transactions A*.

One may wonder why *PRL* immediately accepted the paper, whereas the first, original manuscript experienced such obstacles. It may be suggested that that has to do with the presence of Cahn on the author list of the *PRL* paper. Referees can be impressed by the great, outsized reputation of a name and thereby adopt a more positive, less critical attitude towards a submitted and to be refereed manuscript. This effect introduces significant bias in the system of "reviewing papers by peers", that is generally adopted as the system of least evil…

It appeared that theoretical considerations had appeared before about atomic arrangements that could give rise to "aperiodic" crystals that yet, exhibit perfect long-range geometrical order but that do not show translational periodicity. One type of "aperiodic" crystals is the one discovered by Shechtman: crystalline material exhibiting "impossible" rotational symmetry. These materials have been called *quasicrystals*.[2]

Also after publication of the *PRL* paper, the criticism did not come to an end. The most notable opponent was Linus Pauling, a two times winner of a Nobel Prize. Pauling did not cease in trying to find a way to explain

---

[2] Another type of aperiodic crystals is the "incommensurately modulated atomic structure": a modulation is superimposed on the parent crystal structure. The modulation can be of compositional nature or of positional nature. If that modulation is incommensurate then it equals an irrational number of a translation period of the underlying parent crystal structure, i.e. x times the translational period of the parent crystal structure can never equal y times the translational period of the modulation. The existence of such incommensurately modulated crystal structures has been reported increasingly since 1950.

the Shechtman observation by sticking to periodic crystals subjected to very special ways of twinning. Once, before a large audience at a meeting, Pauling stated: "Danny Shechtman is talking nonsense, there are not quasicrystals, just quasiscientists." That hurts.

Linus Pauling died in 1994. In 2011 Shechtman was awarded the unshared Nobel Prize in chemistry for his discovery of the quasicrystals. It was the crowning conclusion of a long fight for recognition for an in the end indisputable experimental finding.

9.2 The Kirkendall Effect; Unequal Elemental Diffusion Rates

Diffusion, as a process to bring about transport of a component in and through a medium, has been a focal point of scientific interest for already a very long time. Apart from scientific curiosity, there is a practical reason for that: material properties are often optimized by inducing desired microstructural changes and such changes require migration of entities as atoms or molecular entities, i.e. diffusion.

For quantitative evaluation of the progress of diffusion, in 1855 Adolf Fick presented his first law as a phenomenological result and derived his second law (see Chap. 4 for the notion law). These laws describe the transport of the component in the medium considered as a function of time and place. The nature of the diffusing component and that of the medium wherein the diffusion takes place are expressed by a parameter called "diffusion coefficient".

Considering diffusion in multicomponent systems it was realized that all components can diffuse. Thus for a binary (=two-component) system A-B, which can be constituted such that a block of pure A is initially attached to a block of pure B, diffusion of A into the initially pure block B takes place simultaneously with diffusion of B into the initially pure block A. Now suppose that A and B are crystalline elements, say metals (e.g. A = copper and B = nickel; a classical system for studying diffusion in solids) and that in the A-B diffusion couple indicated above diffusion occurs; one then speaks of *inter*diffusion. Upon interdiffusion the atoms A are likely to occupy sites on the crystal structures of A and B and similarly for the B atoms. The diffusion thus takes place on the crystal-structure sites (i.e. the atoms do not take permanent positions in-between); this is called substitutional (i.e. A replaces B and vice versa) diffusion. Then the simplest diffusion mechanism that one can imagine is that (neighbouring) A and B atoms exchange positions (on the crystal structures) upon interdiffusion: the direct exchange mechanism (a particular variant of this mechanism is the so-called ring mechanism, where cooperative migration of atoms is realized by a "ring" of atoms). As a consequence of this diffusion mechanism the diffusion rates/the diffusion coefficients of A and B must be equal. This was the picture that was held until late in the fourties of the past century.

Ernest O. Kirkendall, at the age of 21, in 1935 started a project that would lead him to a doctor's degree at the University of Michigan. His topic was the diffusion taking place in brass, i.e. a copper-zinc (Cu–Zn) alloy. The diffusion couple consisted of brass electroplated with copper. During this research Kirkendall noticed that the original interface of the diffusion couple had moved away from its original position. This phenomenon is incompatible with equal values for the diffusion coefficients of copper and zinc; i.e. the observation cannot at all be reconciled with a direct exchange (or ring) mechanism of diffusion. His promotor and supervisor made it clear to him that he would not get his doctor's degree if he insisted on an interpretation that rejected the direct exchange mechanism. Kirkendall defended his thesis successfully in 1938 and in his first paper published in 1939, with his promotor/supervisor as co-author, there was no indication of a conclusion that the diffusivities of copper and zinc had to be unequal. Kirkendall, at the age of 25, had been forced to bow for the ruling paradigm in diffusion science.

Next Kirkendall was appointed at the Wayne University. He repeated his experiments on the brass-copper system and wrote a second paper, published in 1942, and now entirely on his own, where he showed, by metallographic analysis, using a light optical microscope, that the original interface had moved and claimed that this could not be explained by a direct exchange (or ring) mechanism.

The definitive experiment was performed some years later and completed after the Second World War. Kirkendall now was joined by a student, Alice D. Smigelskas. They devised the now famous experiment where inert, insoluble molybdenum (Mo) wires were placed at the original interface (the interface where the brass and the copper were bonded), as markers, in the Cu–Zn alloy/Cu diffusion couple, to demonstrate unambiguously the migration of the original interface upon interdiffusion. It was concluded that the diffusion coefficients of copper and zinc were unequal: zinc diffusing faster in copper than copper in the copper-zinc alloy.

The paper was submitted to the *Transactions of the American Institute of Mining and Metallurgical Engineers* (*Trans. AIME*) in April 1946 and rather immediately rejected by Robert F. Mehl. Mehl was an important, influential materials (metals) scientist and in particular a world leading scientist in the field of diffusion in metals. He could not believe the results obtained by Kirkendall. Others convinced Mehl eventually to let Kirkendall present his work in November 1946 at a meeting of the Institute of Metals Division; Mehl and others could then put forward their remarks and criticisms in the discussion of the paper that was also printed immediately after the paper.[3] And so it happened: both the printed paper and its discussion were published in *Trans. AIME (171 (1947), 130–142)*.

The paper invoked enormous resonance: the discussion part comprises eight pages and thereby is longer than the paper itself. Mehl now not directly criticized the work by Smigelskas and Kirkendall, but remarked that previous references in the earlier literature claiming faster diffusion of A into B than of B into A were shown to be wrong and remarked that if Smigelskas and Kirkendall would have right, then that would mean that two diffusion coefficients were needed to describe substitutional diffusion in binary solids and *that the solution to the diffusion equation under these circumstances would be extraordinarily complex.* Clearly Mehl did think, and at least hoped, that the Smigelskas and Kirkendall work was wrong.

In fact, the work by Smigelskas and Kirkendall implies that, not an exchange mechanism, but a vacancy mechanism for diffusion describes substitutional diffusion (i.e. an atom A does not exchange crystal-structure site with an atom B on a neighbouring crystal-structure site, but the atom A jumps to a neighbouring *empty* crystal-structure site (=vacancy)). This consequence was not mentioned explicitly in the Smigelskas and Kirkendall paper. Remarkably, during the discussion mentioned above, it was a co-worker of Mehl, Roman Smoluchowski, not afraid of taken a position adverse to that of his

---

[3] In these days discussion of oral presentations at a meeting were taken much more serious than nowadays. In present-day meetings five minutes per oral presentation may be allotted for discussion of work presented. In that way no significant exchanges of opinions can occur. This is regrettable, as progress of science depends on discussion, especially in case of controversial results as is the case here.

In 1981 I attended the conference "Heat Treatment '81", organized by the The Heat Treatment Committee of The Metals Society (Birmingham, UK). This was the first and last conference, that I attended, where not only lengthy discussion time was reserved but also the discussion was recorded and typed in shorthand and then, of course, taken up, in edited form, in the published proceedings of the meeting. Even now, rereading these published discussions is not only a pleasure but also very informative and promoting scientific understanding, also because participants in the discussion used the occasion to briefly present own (preliminary) results of relevance to the discussed theme (precisely this is also the case in the discussions of the Kirkendall/Smigelskas and Darken papers dealt with above). One may suspect on the one hand, that the costs of preservation of such discussions nowadays are considered as too high, but, on the other hand, it may very well be that many presentations at nowadays conferences are considered to be too unsubstantial and/or of a too low quality, that substantial discussion would be worthwhile (see Chap. 6, where it is argued that the average quality of nowadays published papers is lower than in the past).

superior, who said that the Smigelskas and Kirkendall paper convincingly(!) demonstrated the role of vacancies in diffusion.

One year later L.S. Darken presented such "*extraordinary complex*" analysis of diffusion (indeed, two diffusion coefficients are needed...), on the basis of the Smigelskas and Kirkendall data, in an other classical paper also published in *Trans AIME (175 (1948), 184–194)*. In the published discussion part, directly following that paper, Mehl and co-workers (now much more negative than in the above referred to direct discussion of the Smigelskas and Kirkendall paper) remarked that Darken's analysis is based on "*one experiment of questionable validity*" (i.e. the experiment performed by Smigelskas and Kirkendall). Mehl et al. then continue and have the following further disqualifying comments: the paper by Smigelskas and Kirkendall "*contains serious inconsistencies*"; "*the shape of the penetration curves reported by Kirkendall and Smigelskas....is suspect*" and "*Taking all data into consideration, there seems to be no reason for accepting the results of the Smigelskas and Kirkendall experiment*". The young Kirkendall must have felt condemned.

It took much time and further results confirming and thereby validating, what is now called, the "Kirkendall effect". Then, at last, also Mehl was convinced of its genuineness. In the meantime Kirkendall had given up his academic career in already 1947 and had accepted an administrative job as secretary of the AIME. At the time he even may not have been aware of his "victory". His short scientific career comprises only three papers. Nowadays his and Smigelskas' celebrated work is cited in every textbook dealing with (also) diffusion.

It can be suggested that the damaging criticism of his work, expressed by high ranking peers, had pushed Kirkendall to abandon pursuing an academic career (about five decades later and being over 80 years old, he said to have left science in 1947 because of financial reasons). In any case: the fierce opposition to his work certainly will have made leaving science, and that forever, attractive.

Mehl, many years after 1947, apologized personally towards Kirkendall for his former unjust, long lasting, fierce opposition. Noteworthy, a similar act of fairness towards Shechtman apparently was beyond Pauling's capacity (cf. Sect. 9.1). Mehl died in 1976, Kirkendall in 2005.

9.3 Clusters or Precipitates

After having published a couple of papers from our nitriding research (cf. Chap. 8), one day in the beginning or the eighties of last century, I out of the blue received a letter from Professor Ken(neth) H. Jack. I had not had contact with him before. Of course Jack's name was known to me: Jack had performed groundbreaking research in the fourties and fifties on the crystallography of interstitial phases, especially iron nitrides and iron carbonitrides. This work was done on the basis of X-ray powder diffraction patterns recorded with photographic techniques. Later, as a Professor at the University of Newcastle upon Tyne (United Kingdom), he maintained this line of research now also studying the nitriding of iron-based alloys.

The letter invited me, together with two other scientists working on "nitriding", Professor Jacques Foct (University of Lille, France) and Professor Donato Firrao (Polytechnic of Turin, Italy), to Newcastle for an informal workshop. I accepted and went to Newcastle after having prepared a presentation of our results.

Jack's group was housed in a former brewery, with his equipment, still *photographic* X-ray diffraction instruments (rather unusual already at that time; diffractometers, allowing direct intensity recording with very high angular resolution, had found much more widespread application) and a high resolution transmission electron microscope, put up in the basement of the old building.

On the one hand Jack was "social" and very active to make the stay comfortable for his visitors; the four of us spent a few very enjoyable dinners and evenings together. On the other hand, in scientific discussions, where I, back then and now, do strongly welcome directness, Jack was more than direct, even tactless and rather blunt. Donato was already accustomed of being accused, by Jack, of telling nonsense and told me about that, which "armed" me in advance of such discussions.

Now some details have to be presented in order to make clear why my controversy with Jack arose. I presented results of our research on the precipitation of an iron nitride, so-called α''-nitride ($Fe_{16}N_2$), in nitrided pure iron (Fe) and of chromium nitride (CrN) in nitrided iron-chromium (Fe–Cr) alloys. In both cases the results showed a.o. that α'' iron-nitride and CrN (nano)*precipitates* had developed in both systems, respectively. Jack did not accept this interpretation. In former years he had developed the concept that precipitation in these systems, nitrided pure iron and nitrided iron-based (Fe-Me; e.g. Me = Cr) alloys, was preceded by clear stages of *"clustering"*. Clustering involves (i) that in such an initial stage of precipitation of nitrided iron (Fe) local enrichments of nitrogen atoms would develop, exhibiting no ordered arrangement of the nitrogen atoms as is the case in a precipitate as the α''-nitride, and (ii) that in such an initial stage of precipitation of MeN (e.g. CrN) in nitrided Fe-Me alloys local enrichments composed of Me and

N occur, exhibiting no ordered arrangement of these atoms as is the case in a precipitate as the chromium nitride. The clusters of type (ii) were called by Jack: *"substitutional-interstitial solute-atom clusters"*.[4]

Jack and co-workers had bombarded the literature in the preceding couple of decades with their interpretation, i.e. their concept of "substitutional-interstitial solute-atom clusters". I was not impressed and stated that no conclusive evidence was ever presented by them; on the contrary, their published data already indicated, and I discussed their claimed (diffraction) evidence, the presence of (nano)precipitates, i.e. no clusters. At this stage the difference in interpretation could have simply been discussed at length and perhaps settled with a common conclusion. However, such an approach was incompatible with Jack's character. Perhaps he had not expected from a much younger man, of distinctly lower rank and status, such opposition and could also therefore not comply with what were hard facts. He took it personal and, as turned out later, it had deteriorated the chance for any possible fruitful contact with him.

A few years later, I, together with Marcel Somers (at the time a Ph.D. student in my group; later a well-known Professor at the Technical University of Denmark (Lyngby) and prominent in the field of nitriding) and Ruud Lankreijer (a Master's student) submitted a paper to the *Philosophical Magazine*. In that work we focused on the quantitative modelling of the uptake of extra nitrogen, so-called "excess nitrogen", in nitrided Fe-Me alloys due to the straining of the parent (iron) crystal structure as a result of the presence of the misfitting MeN precipitates. In the paper iron alloys with Me =

---

[4] To explain this name: the Me (e.g. Cr) atoms reside on sites of the parent (iron) crystal structure (i.e. they are *substitutional solute* atoms); the N atoms reside on interstitial sites of the parent iron crystal structure (i.e. between the sites, on interstices, of the parent (iron) crystal structure; i.e. the N atoms are *interstitial solute* atoms).

It may be argued that any precipitation process of a compound in a supersaturated matrix begins with the transport of dissolved atoms to the site where the precipitation is to be initiated. For some time an enrichment of these dissolved atoms on sites of the parent crystal may exist before a precipitate particle with its crystal structure different from the parent matrix develops. These enrichments, devoid of long range ordering and therefore called "clusters", in some, rather rare cases, can exist for a considerable period of time. For example, they have been proven to occur in aluminium-copper (Al–Cu) and aluminium-magnesium (Al–Mg) alloys. These clusters have been called Guinier–Preston (GP) zones after their discoverers. However, in many systems no cluster stage is observed at all; more or less immediately extremely small nanoparticles with the crystal structure of the precipitating compound, and no clusters/enrichments of dissolved atoms on sites of the parent crystal structure, are detected. In case of the precipitation of α″-nitride in iron supersaturated with nitrogen, clustering of nitrogen is (already) unlikely to occur, as the nitrogen atoms, moving from interstitial site to interstitial site, arriving at the site of precipitation have only to occupy specific interstitial sites in order to realize the α″-nitride crystal structure, in association with pronounced relieve of (strain) energy. Indeed, no evidence for a distinct clustering stage in this system has ever been presented. On the contrary, in the very first stage of precipitation the crystalline α″-nitride has been identified (see also "*Iron Nitrides and Iron Carbides; Tempering of Steels and Pre-Precipitation Processes*" in Chap. 10).

Ti (titanium), V (vanadium), Cr (chromium) and Mo (molybdenum) were considered; all systems also investigated in the past by Jack.... The editor, as was also unmistakably clear from the resulting referee report, sent the manuscript to Jack for review. After considerable time we got the referee report with the conclusion of the referee that our paper was entirely wrong, had no merit and that "*It is doubtful whether any useful purpose could be served by recommending changes......*". This referee report had a length of about 20 densely typed pages.[5] I had never seen before and have not seen thereafter a referee report of such length. The referee (Jack) did not agree with the presence of (nano)precipitates; there had to be clusters and therefore our approach was worthless. After ample consideration I and Marcel decided to write an extensive rebuttal dealing with all points and arguments raised by Jack and requesting the editor to reexamine the case. In the cover letter we indicated that it would not be wise to consult Jack again as he was obviously strongly biased. We also added to the paper two appendices where we listed and discussed all data and results used and interpreted erroneously by Jack et al. in favour of the cluster interpretation and demonstrated this interpretation to be untenable and that instead the *available* data support the (nano)precipitate interpretation. The editor complied and the paper was accepted for publication at last.

At about the time of the above described "paper fight", I attended a conference on nitriding and related phenomena and processes in London, organized on behalf of The Metals Society by Professor Tom Bell (University of Birmingham). Tom was also very active in nitriding research. He had a personality completely different from that of Ken Jack. He was very open to any novelty in his field of science and liked to be surrounded by young, "upcoming" scientists. It was told he had had similar problems with Jack as I had. Soon thereafter he made me co-editor of a journal in our field.

Jack was also attending the meeting. On one occasion he went up a staircase, at the moment that I went down on the same staircase. We passed along each other. I greeted him, as I still thought that a scientific difference of opinion should not be an insurmountable burden for a normal, personal relationship. Jack deliberately looked away and did not say anything.

Both Jack and I were later invited to present main lectures at a conference in Poland. At that time Jack was already retired and for many years toured the world presenting lectures about his work from the past. Of course he never neither mentioned orally nor referred in the one or the other review paper, which he wrote, to our work. At the conference dinner and party he drank

---

[5] I am grateful to Marcel Somers who kept a copy of the referee report; the original got lost upon my move to Stuttgart in 1998.

too much. I brought him to his room and put him in bed. The next day he came to me and warmly thanked me for taking care of him. I thought this could be the start for improving our relationship. However, I would be taught otherwise.

I had moved to Stuttgart in 1998 and had originally intended no longer to investigate "nitriding". However, as the result of a particular experience (see *"The discovery of amorphous precipitates in a crystalline solid"* in Chap. 13), after a few years I decided to restart research in the area of nitriding and now on an even more encompassing scale than in Delft. We published a lot. Jack was now close to 90 and I did not expect him to be active in some way or even to follow what we presented. However, one day one of my Ph.D. students showed me a (I seem to remember handwritten) letter of Jack to my student commenting on his nitriding work; Jack tried to "sell" again his concept of "substitutional-interstitial solute-atom clusters". In no way Jack in this letter made any reference to me; I forever remained a "fiend" for him.

Jack died in 2013, 95 years old.

10

On Burgers' Chair

Abstract A few notes on the scientific life of Professor Wilhelm (Willy) Burgers and the significant role of his brother Professor Johannes (Jan) Burgers. My personal reminiscences of Burgers. Description of the Chair as led by me after 1985 and its position within the Laboratory. Examples of major research projects: (i) the determination of the structural imperfection in crystalline materials, by the analysis of X-ray diffraction line profiles, highlighted by the development of the so-called single-line Voigt method, and (ii) the *pre*-precipitation processes in steels; these narratives have been chosen also because they as well allow indication of experiences of scientific behaviour of questionable decency. A visit to colleagues in the former DDR, shortly after the "Wende". Unjustly overlooked scientists. The growing importance of (ever more) evaluations in science. Why I left "Delft" for "Stuttgart" in 1997/1998.

Professor Burgers

One way to describe the research work by Wilhelm (Willy) Gerard Burgers is to mention his interest in the structure of crystals and especially their defects.

In 1918, at the age of 21, as a student of the University of Leiden, Burgers attended a couple of the famous Monday morning lectures by Lorentz (then as extraordinary Professor, after his move to Teylers Museum in Haarlem; see, in Chap. 3, the intermezzo "*On Lorentz and Points of Contact; the Lorentz Factor*"), which dealt in particular with the recently determined crystal structures of salts like rock salt (sodium chloride (NaCl)) on the basis of X-ray diffraction analysis by father and son Bragg. Thereby Burgers was made clear that crystals were composed of atoms *arranged in a regular way*: the atoms in crystals are positioned at constant distances with respect to each other,

© The Author(s), under exclusive license to Springer Nature
Switzerland AG 2022
E. J. Mittemeijer, *How Science Runs*,
https://doi.org/10.1007/978-3-030-90095-3_10

the crystal has an internal regularity and thus can be conceived as a three-dimensional grating (see"*Diffraction Analysis as a Tool for Determining the Constitution of Matter; Regularities and Irregularities of the Atomic Arrangement*" in Chap. 5). That more or less was the ignition point of his following study and career.

In Leiden, Ehrenfest, the successor of Lorentz, took an almost paternal and certainly friendly authoritarian position towards Willy Burgers (and his brother Johannes (Jan), M. Burgers who also studied in Leiden; see further below), as was also his attitude to the other students under his guard. Ehrenfest, in response to the "*Laue-Bragg Dinge*" (= "Laue-Bragg things"), proposed a certain study program to Burgers that ended with the words: "*und dies alles sicher unverheiratet und womöglich unverlobt*" (= and all this absolutely unmarried and if possible unengaged). This was meant likely more serious than funny.

Nowadays a Professor addressing a student in this way might be "sued" as intruding unjustified and unlawfully in private life and obstructing realization of a "life-work balance". Yet, as a Professor, I myself did expect that my students did not consider a Ph.D. project as an assignment that could be done between 9 a.m. and 5 p.m. and only from Monday till Friday (and I told them before the start of the Ph.D. project) and I did consider founding a family with children during the time period of a Ph.D. project as less wise (but did not tell them that explicitly…).

After his Master study in chemistry, completed at the University of Groningen, Burgers went to London to the laboratory of Sir William Bragg (father Bragg), "*as a beginner in research*", as he writes himself. The research he performed their led to his thesis that he defended after his return to Groningen in 1928. At about this time Burgers acquired a position at the Philips Research Laboratories (= NatLab; see what I wrote about the NatLab in Chap. 5) with a focus on X-ray diffraction analysis of metallic materials. Here he showed an enormous drive and developed his great expertise on the recrystallization of metals, the major topic of his scientific career.[1]

In 1934 E. Orowan, G.I. Taylor and M. Polanyi independently introduced the "edge" dislocation as a crystal-structure defect (see Footnote 8 in Chap. 8 and Footnote 2 below) to explain that the deformation of a metal crystal was much easier to realize than could possibly be expected if the atoms in

---

[1] In long ago days strong deformation of metals was thought to destroy their crystalline nature. By subsequent heating the crystallinity would be restored. This background explains the name "recrystallization". We now know that "recrystallization" means the emergence of a change of the crystal orientations of the whole polycrystalline specimen by the growth of new, strain-free crystals ultimately encompassing the entire piece of material considered.

the crystal were perfectly arranged as prescribed by the known crystal structure. Willy Burgers was not such mathematically gifted as his elder brother, Jan Burgers, a from experimental-to-theoretical turned physicist, who was appointed at the age of 23 (!) as Professor of Aerodynamics at the Delft University of Technology. Willy asked Jan for help in order to understand the Orowan–Taylor–Polanyi work. The interest of Jan was aroused and as a final result Jan published in 1939 and 1940 important work that introduced (i) the concepts of the "screw" dislocation and of the famous Burgers vector,[2] that characterizes the displacement introduced by a dislocation in a crystal, and (ii) the concept of low angle boundaries between adjacent crystals as "sets of parallel dislocation lines". Jan Burgers did not stay in this field of science and, instead, became an authority in the field of fluid dynamics. In 1955 Jan Burgers left The Netherlands and went to the University of Maryland in the USA. Remarkably: until the present day the two Jan Burgers papers from 1939 and 1940 on dislocations remain his two most cited papers.

In a sense there is some tragic in the recognition that the Burgers vector is the achievement of Jan and not of Willy. Willy stayed in the field of metallurgy and, consequently, to his embarrassment, was (and still is) often considered as the "father" of the Burgers vector. I was told that on one occasion, where a honorary doctorate was granted to Willy Burgers, the French speaking eulogist mentioned that science owed the Burgers vector to Willy Burgers. Burgers immediately interrupted and shouted: *c'est le vecteur de mon frère, c'est le vecteur de mon frère!!* (= it is my brother's vector, it is my brother's vector!!).

The narrative about the emergence of the Burgers vector, by Willy Burgers, in his very last paper, with the title "*How my brother and I became interested in dislocations*" and published in 1980, is very worthwhile to read (*Proceedings of the Royal Society (London), A371 (1980), 125–130*). It makes clear that Willy was most pleased with his role as the "uncle" of the Burgers vector.

In 1940 the appointment of Burgers as Professor in the Faculty of Chemical Technology of the Delft University of Technology followed. He built up a group where X-ray diffraction and electron diffraction and microscopy became the major tools for microstructural investigation; he maintained the line of research he had followed before.

---

[2] If the direction of the Burgers vector is perpendicular to the direction of the dislocation line (see Footnote 8 in Chap. 8), one speaks of an "edge" dislocation. If the direction of the Burgers vector is parallel to the direction of the dislocation line, one speaks of a "screw" dislocation. The characters of the distortion fields surrounding the dislocation lines are essentially different for "edge" and "screw" dislocations. Dislocation lines need not be straight, they can be curved and then they have no constant line direction. However, the direction of the Burgers vector for any dislocation is constant. Thus a curved dislocation line is of variable, "mixed edge-screw" character.

He remained a front runner in the research on recrystallization. An important controversy raged in the literature on recrystallization regarding the origin of recrystallization. As a Ph.D. student I became aware of that, especially because a former Ph.D. student of Burgers, Professor C.A. Verbraak at the Twente University of Technology, continued research in this area; a Ph.D. thesis of his group in 1977 had the title "Oriented nucleation or oriented growth" (by J.W.H.G. Slakhorst), which dealt extensively with the opposing views indicated in this title. Only in the years after the death of Burgers it became more or less established that subgrains/dislocation cells, resulting by the deformation process preceding the recrystallization, induced by subsequent annealing become the origins of the recrystallization (see my book *"Fundamentals of Materials Science"*, Second Edition, Springer, 2021, pp. 589–598).

In 1941 Burger's book on recrystallization appeared: *"Rekristallisation, verformter Zustand und Erholung"* (= *Recrystallisation, Deformed State and Recovery*) (Handbuch der Metallphysik, vol. 3, pt. 2, Akademischer Verlaggesellschaft, Becker & Erler Kom.-Ges., Leipzig, 1941). This voluminous work was the first ever monograph on recrystallization. It was written in German, which was an obstacle for dissemination of its contents in the English speaking world and regrettably has led to irritatingly erroneous "citations" of what Burgers has stated in this book. (Within this context see the text under "*German versus, the lingua franca of science, English*" in Chap. 8).

Other topics of importance dealt with in the Burgers group were diffusion in metals and phase transformations. Burgers retired in 1967.

The publication pressure of those days was much smaller than today (see Chap. 6). In that light we should look at the following data (almost all taken from the Web of Science (WoS) and therefore they represent underestimates, as the WoS does not cover all published work). At the end of his career Burgers had (co-)authored somewhat less than 200 papers (the WoS lists 114 of these, suggesting the nonlisted papers may have appeared in lesser known journals. Indeed, a substantial part of those papers not covered by WoS concerns papers in Dutch which appeared in local, Dutch journals). His most cited paper is from 1934 (i.e. from his time at the Philips Research Laboratories (NatLab)) on the martensitic (called "homogeneous" by Burgers) transformation of zirconium (Zr), of which he is the sole author: the paper has drawn until now more than 1100 citations, with in the very recent years 2018–2020 an average of 70–80 citations per year, which is remarkable for a paper published about 90 years ago. He has been the supervisor of 25 Ph.D. theses.

The structure of the group as developed by Burgers, with a focus on X-ray diffraction and electron diffraction and electron microscopical methods for microstructure characterization, was largely upheld by his short- till very short-period successors Okkerse and Radelaar. Also I maintained this structure. Therefore in this booklet I speak of the "Burgers Chair", a notion with resonance in, at least, The Netherlands.

My contact with Burgers

The first time I saw and "experienced" Burgers was by following his lectures on Physical Chemistry of the Solid State, which he gave, as one of the "capita selecta", a last time in 1970/1971 in the Laboratory of Metallurgy, as Okkerse, just appointed, did not feel or was in the position to take care of it already. I summarized elsewhere my memories of the sensation felt during these lectures and cannot do better than repeat them here.

At the time, he was of course a rather old man, but, while lecturing, still had the enthusiasm and dynamic aura of a young man (which was the more remarkable in view of his physical handicap), presenting in this lecture course basic, established knowledge, interspersed with anecdotes and research results obtained by others and himself, and in particular thereby was capable to transfer to the audience his love for science. This lecture course has brought about my decision to turn to materials science, for my Master's project and, later, my Ph.D. project, not realizing that it would be the playing field for my whole career.

Burgers had lost one leg (a consequence of an accident? I do not remember and I never felt in the position to ask him). The artificial replacement allowed slow, tedious walking accompanied by slight, creaking sounds; the device needed regular adjustments and/or repair by a technician of the group. This was certainly very unpleasant for Burgers. However, much worse was his suffering from phantom pains, as I learned much later. These pains, which give the impression that they originate from the amputated body part that no longer is there, are a response from the nervous system (the brain). As a remedy Burgers took a really strong pain killer. Nothing of this suffering he showed.

Burgers was a man who kept distance, which did not hinder the development of a good relationship. But in this sense he was a "classical" Professor. I certainly did not belong to those close to him. During my Ph.D. project, under the supervision of Okkerse, I once asked him for advice on the sodium chloride/potassium chloride (NaCl/KCl) system, since we planned to extend our diffusion project from metals to salts and it was known that under Burgers some research (a Ph.D. thesis in Dutch had remained of that work), had been carried out with this system. On another occasion he mediated in an understanding way in a minor conflict about priority I had with a former Ph.D. student of him. And, finally, as mentioned before, he was a member of the committee for the public examination of my thesis.

I believe Burgers became more seriously "interested" in me only after I was nominated on "his" Chair. Burgers, as emeritus, had a small office in the Laboratory. Here we spoke a few times with each other, also, perhaps especially in small talk, on topics besides the daily work, as about previous times and his (former) contacts with personalities of names well known to me. Thus I remember he told me that he had once visited the Max Planck Institute for Metals Research in Stuttgart in the time it was led by Werner Köster, a scientist he appeared to know well. Not at all did I realize at that time that this institute would be my affiliation in the second half of my career, where I would occupy, next to a Directorship at the Max Planck Institute, a university Chair once occupied by Köster.

Burger's office was full of ping-pong ball models, which represented a great "hobby" of his. They played a large role for demonstrations in his lectures. Therefore I chose the photo shown here for a picture of Burgers (Fig. 10.1). Those of my readers a little familiar with materials science will immediately recognize the model in front of Burgers as representing the symmetrical, small angle tilt boundary as a series of parallel edge dislocations at constant distances; by the way, as originally proposed by his brother in 1940…

A copy of his "recrystallization" book lay on his desk. He told me that it had been an enormous, thwarting burden to finish this book next to his daily tasks; it had almost "killed" him (his words). I would have to share this experience, years later, while drafting my *"Fundamentals of Materials Science"* albeit in active duty.

One of the awards he got was the Acta Metallurgica gold medal. Burgers could be humorous. Thus he told me chuckling that he had measured the volume of the medal and its mass and thereby had determined its effective density. On this basis, he came to the conclusion that the medal was not at all made of pure gold (alloy). The effective density was compatible with a model of the medal as a thin case of gold (alloy) likely filled with sand.

Burgers died in 1988 at the age of 91. On behalf of the faculty I spoke at his funeral. His wife had died some years earlier. They had no children. His brother had died as well, in the USA. Thus it came that somewhat remote members of the family had the task to empty the spacious house in Rijswijk, not far from Delft, where Burgers had lived. I was asked to come to the house. The family members had no connection with the "world" of Burgers; moreover, they had not at all a natural science background. That made it difficult for them to see the worth and meaning of much what had been

Fig. 10.1 Prof. Willy G. Burgers in his office. In front of him on his office desk (i) at his right side (for the observer at the left side of the photograph) a cube of pingpong balls in the cubic closest packing of identical spheres and (ii) at his left side (for the observer at the right side of the photograph) a pingpong ball model of a so-called small angle symmetrical tilt boundary, as originally proposed by his brother Jan M. Burgers. In his hands Burgers holds a sheet of recrystallized aluminium (the large crystals can be observed individually). It is one of such sheets which, many years after the death of Burgers, hung on a wall in my office at the Max Planck Institute and served as background image for the cover of this book (see main text and Fig. 13.6) (taken from *"De Technische Hogeschool te Delft 1905–1955"*, *Staatsdrukkerij-en Uitgeverijbedrijf, 's-Gravenhage', 1955; photograph by Paul Huf*)

stored and kept by Burgers as valuable and meaningful. I was asked to take with me that correspondence that I considered as of some importance. I thus have still in my possession letters to Burgers from scientists who have a name in my field of science, as (in alphabetical order) S. Amelinckx, J.M. Bijvoet, Lawrence Bragg (son Bragg), J. Burke, R.W. Cahn, B. Chalmers, J.W. Christian, J. Friedel, J. Grewen, A. Guinier, R.W.K. Honeycombe, W. Köster, G.V. Kurdjumov, F.R.N. Nabarro, Z. Nishiyama, J. Nutting, A. Seeger, G. Wasserman, C. M. Wayman, and more. This list of names reads as a "Who's Who" in materials science… Whereas a large part of these letters are restricted to transferring of congratulations, also scientific discussions are part of these correspondences.[3] If I didn't know it already, then it was clear now that Burgers had been a very good "networker".

In his home office Burgers had made a small exhibition of the medals and awards given to him in his long career. The family members had no interest at all and I could take a few of these with me. There was however one exception: the Acta Metallurgica gold medal. The family members planned to sell it as they thought it was made of pure gold. I have to admit that I did not make them wiser; for me selling of this object felt as disrespectful to Burgers and his lifetime achievement.

On the wall along the staircase in his house Burgers had put five rectangular sheets of aluminium plate (about 30 cm × 15 cm), which sheets had experienced different degrees of deformation and subsequent recrystallization. Each crystal (called "grain" in materials science) in a sheet was that large that it could be seen by the naked eye. After etching, the grains in a sheet reflected the light differently. This in principle gave a pleasing, visual image, different for each sheet. I was allowed to take the sheets with me. In the state they were they needed new frames and surface cleaning. For more than the next 10 years I did nothing with these sheets; I simply stored them. But I took them with me to Stuttgart. A technician there made new frames, cleaned and newly etched them. Thereafter they were as they must have been originally; their variable, silvery reflectance made a beautiful impression (also see the sheet of recrystallized aluminium in the hands of Burgers in Fig. 10.1). I put them on the wall in my office at the Max Planck Institute (see Fig. 13.6).

---

[3] According to my judgement, these letters have no particular significance for understanding details of the progress of materials/metals science in a passed time. This differs with previous times where science proceeded also by exchange of handwritten or typed letters between scientists (e.g. see the extensive correspondence of H.A. Lorentz). Nowadays we only rarely write letters; we send each other usually less well considered emails, possibly with attachments. These electronic "documents" are only seldom preserved. This will make it more difficult for future historians to retrieve how science developed also by non-public discussion between colleagues and by non-public debates between adversaries.

Often visitors asked me what it was. I used them as illustration of recrystallisation in lectures for students and pointed to the sheets during examinations in my office when I had decided to ask a student something about recrystallization. Finally, a photograph made of one of these sheets served as background for the cover of my book "*Fundamentals of Materials Science*" and now also, in a different fashion, for the present book; thereby the essence of materials science, property change by microstructural manipulation, is well illustrated. I believe that Burgers would be pleased to see how I have dealt with his legacy.

The group

My appointment in 1985 as Professor of Physical Chemistry of the Solid State meant I became the head of an existing group, which I knew well, because I had performed my Ph.D. project there and had maintained cooperation with the group during my time in Korevaar's group, already only because my research also depended on the accessibility of X-ray diffraction techniques and transmission electron microscopy, which were the main experimental tools for microstructural analysis. This description suggests that I had a facile start; no substantial time was needed to settle in.

There was also a likely more difficult aspect connected with this for me new situation. The senior staff members were older than I was and had known and guided me as a student. However I was accepted and no serious difficulties of the kind hinted at arose. The atmosphere in the group during the years I was responsible was generally very good and the group became strongly productive and was successful.

At this place one complication connected with my start on "Burgers Chair" should be mentioned. The most senior scientific member of the group had had strong hopes for being appointed on the Chair. On the day that I was informed that the nomination committee proposed me, he got the message that he would be passed over. By chance we passed each other on the corridor. He looked away and also I couldn't really conceal my embarrassment. It was a most painful situation. He must certainly have been very disappointed and stayed away from the Laboratory for a couple of days. Soon I sought a discussion with him about the situation arisen. I told him that I didn't have the intention to actively interfere with his research and that he could keep his relative independence as long as he stayed loyal with the group. In fact his research work, around the atomic structure of grain boundaries, was not of my immediate research interest. Indeed we never jointly appeared as co-authors on a paper. He had a gentle character, he was congenial and he never opposed me in any serious way. In fact, when we had to fight hard to acquire funds for high resolution transmission electron microscopes we worked well and successfully together (see at the end of this chapter).

A (sub)group was added to the Chair with a focus on electron probe microanalysis, Auger electron spectroscopy and X-ray photoelectron spectroscopy. These methods allow the quantitative determination of composition(-depth) profiles in a specimen as well as the composition at the surface of specimens, which possibilities were of great importance for my own research. The older scientist, who, in the first years of my responsibility, was in charge of this equipment, Daan Schalkoord, largely focused on method development, i.e. bringing these methods to a higher level, which was well compatible with the mere service task of this subgroup. He was the oldest scientist of us all and had to retire in a few years. Two experiences connected with this co-worker are worthwhile to tell here.

Many requests for composition(-depth) analyses from especially other groups were handled by Daan. On one occasion such data obtained by Daan, and given by him to the scientist concerned of another, engineering oriented group in the Laboratory, were taken up into a publication. By itself nothing unusual and in fact a good thing for the Laboratory, and for the service group, as this contributed to the basis of existence of the group. However, Daan checked the resulting publication as published in a journal and found out that his data were unjustifiably and misleadingly "cleaned": data points which did not support the desired conclusion of the authors were simply left out. Daan was upset; me too. This was the first and only time that I was confronted with blunt data manipulation in my research environment.

In view of his age Daan had intensely experienced the Second World War and the occupation by the Germans. He once said to me that he never would make a travel to or across Germany; if he had to go to Poland, for example, he made a large detour. I do not know what has happened to him or his family or whatever could have been the origin and explanation for his outspoken attitude, but it likely must have been something drastically touching. I was surprised. I had many contacts with Germany and Germans, especially because of my work on nitriding, and could not imagine that more than 40 years after the War such general condemnation was possible. I now wonder what Daan thoughts were once he heard much later (I don't know if that really happened) that I would leave Delft permanently for Stuttgart and would even, still more later, adopt the German nationality next to my Dutch nationality.

We cannot and should not forget the especially rather immediate past. For the older generation in The Netherlands the ever present feelings of subordination of a small country with respect to its much larger neighbour[4] were added to the occupation experiences. This can make understandable the reaction described above. Now, 75 years after the Second World War, I have the impression that the current after war generations are not essentially affected such that they approach each other with pronounced reservations on the basis of what is now, albeit recent, history. The dream of a European supranationality has substance.

The Chair thus was composed of three subgroups: one largely involved with X-ray diffraction (XRD) techniques, one utilizing transmission electron microscopical (TEM) techniques and the (surface analytical) service group described above. The XRD and TEM subgroups were largely performing research within the research program of the Chair but they also provided substantial service to projects from other groups.

The team of the Chair was thereby rather large as compared to other groups within the Laboratory. This constituted a problem growing with the years I was in charge, because, under the severe savings program imposed by the university on the personnel budget, of in particular the tenured positions (scientists and technicians), it became increasingly difficult to fill occurring vacancies. Especially the other, smaller groups in the Laboratory and the Faculty had the inclination to look to the largest group for accommodating the required reduction. Consequently substantial energy was spent on repelling such "assaults" and preventing them, also by drafting reports exposing the substantial service loads originating from assignments by third parties. I was largely successful there, certainly also because of our very good reputation. However, I saw that in the end this pressure could not be blocked and this was one, although minor, driving force for me to later accept the available Director position at the Max Planck Institute in Stuttgart.

The present-day situation at a university, that even a minimal number of Ph.D. student and post-doc positions for a Chair can only be acquired by writing applications to external funds with success ratios under 30% or even far less, is an illustration of the lack of sufficient permanent funding of universities. I will come back to this point later (also see Chap. 7).

---

[4] Thus I explain the sometimes "special" atmosphere during Dutch-German soccer matches. After the finish of the semi-final match during the European Championship 1988 in Germany, where the Dutch beat the Germans, a Dutch player, showed his bare ass to the German audience. A detestable gesture, that, however, did not originate in the memory of the Second World War, that all Dutch players of course did not experience, but in the joy of the rare occurrence that the soccer team of a small country besieged the soccer team of its large neighbour, that moreover was one of the greatest powers in soccer worldwide. Often I have been confronted with this moment in soccer history with the comment that this hints at a relationship of the Dutch with the Germans that is distorted as a consequence of the German 1940–1945 occupation. I don't think so.

On the same, second floor of the Laboratory of Materials Science also the group led by Professor Arie van den Beukel, as successor of Professor Paul Penning (see "*From Science to Management; the Career of Professor Okkerse*" in Chap. 5), had its offices and laboratories. Both Chairs together constituted the part of the Laboratory where fundamental research was carried out; the other groups in the Laboratory were more engineering/application oriented. In the organizational structure of the Laboratory we formed a so-called "vakgroep" (i.e. a group of Chairs of related character) with the name "Fysische Chemische Materiaalkunde" (= Physical Chemical Materials Science). We acted in a more or less united way within the Laboratory. At the same time, within the "vakgroep" we regularly had long and tiring (also see the preceding paragraph) meetings regarding issues where strong difference of opinion between the two Chairs could occur, but eventually acceptable compromises always resulted.

Van den Beukel had spent his entire career at the university in the group he now headed. He obtained his Professor rank without being nominated by a selection committee but by the government decision of 1980 to abolish all "lectores" positions (a "lector" position was something like a "reader" position at UK universities) and to confer the title Professor to all previous "lectores". Arie had a sharp mind and a sharp tongue. He loved smoking cigars; strikingly, he inhaled the smoke, as normally done by cigarette smokers. Over the years I noticed his facial skin obtaining a more parchment appearance, a characteristic of strong smoking.

I did respect him and this was also the reason that I had approached him in 1977, at the end of my Ph.D. project, with the request to give me his opinion about which offered position to prefer: the one in Korevaar's group or the one at the Philips Research Laboratories (see at the end of Chap. 5). He strongly advised me the latter job. I did the opposite, because I saw the possibilities for my own development in Korevaar's group. The time thereafter proved me right and I wonder if Arie ever remembered our discussion of years ago. I preferred not to ask him.

The field of research covered by van den Beukel was not large, but the quality of the work was very good. He retired in 1998, i.e. at about the same time I left Delft for Stuttgart. During the last 10 years or so of his career his interest was strongly attracted to themes connected with science but not within science. He is a strictly religious man. With that background he felt irritated by scientists having the attitude that God was unnecessary, say superfluous as a concept, in view of the essence and progress of science (science is mostly reduced to physics in the eyes of van den Beukel). He also advocates the opinion that there was insufficient scientific proof for the evolution theory, and attacked the work by Richard Dawkins, Francis Crick, etc. This is what he considers to be the arrogance of the scientists. He published a number of (such) books. The best known probably is his first one: *"De dingen hebben hun geheim; gedachten over natuurkunde, mens en god"* (*Ten Have, Baarn, 1990*; followed by numerous reprints), which also appeared in English, with a somewhat different title, as *"More things in heaven and earth: God and the scientists", London, SMC Press, 1991*). The book is very well written and enjoyable to read, also by somebody who does not share the basic convictions of Arie, which holds for me. After all, there is no need for a discussion about the issues raised by van den Beukel: in the end the progress of science will deliver answers or not, and there is nothing that we can do about that (see the end of Chap. 4).

Diffraction line profile analysis; the Voigt function and the "Dream Team"

A (an X-ray) diffraction peak recorded from a (poly)crystalline specimen, is not infinitely sharp: it is broadened by the smallness of the size of the crystals constituting the material and by mistakes, as dislocations, in the ideally regular, periodic, atomic arrangement (see "*Diffraction Analysis as a Tool for Determining the Constitution of Matter;…*" in Chap. 5). Moreover, instrumental effects, as owing to the width of slits through which the X-ray beam passes and the non-monochromatic nature of the X-rays used, also lead to broadening of the observed diffraction peak (also called diffraction line *profile*).

The materials scientist needs the microstructural information, as indicated by the crystallite size and mistake density, to understand and predict the (for example, mechanical) properties of a material. Hence, in the profile analysis, after the profile measurement, it is necessary to separate the instrumental broadening from the physical broadening and to separate the physical broadening effects due to the individual microstructural parameters. This activity is called "*diffraction line profile analysis*". The many decades passed since the discovery of X-ray diffraction have seen the proposals of very many methods to perform this task. Most of these are primitive or problematic as they suffer from assumptions invalidating the physics of the diffraction process and/or lead to data of unclear interpretation. This, at least, describes the impression I had when I started my Ph.D. project in the seventies.

Thus, more or less "forced", we, i.e. Th.H. (called: Staf) de Keijser, Rob Delhez and I, started in Delft research on "diffraction line profile analysis".

It would become an activity that accompanied me, for a significant fraction of time, till my departure to Stuttgart, but also in Stuttgart I now and then returned to this type of methodological research. It certainly is beyond the scope of this booklet to detail the progress made by us in this area. However, I want to focus briefly here on a special aspect, which is of pronounced usefulness till present day and the history of which is another illustration of how science runs.

In order to realize an analysis of the peak broadening a mathematical function to realistically describe the peak shape is desired. In the field of diffraction analysis quite a number of such functions have been proposed. The first proposed has been a triangle (really)......, which certainly is generally very unsuitable. Later bell-shaped functions as the so-called Gaussian function and the so-called Cauchy function (also called Lorentz function) have been applied. In 1978 J.I. (Ian) Langford proposed the Voigt function to describe the shape of diffraction peaks (*Journal of Applied Crystallography, 11 (1978), 10–14*). A Voigt function is the convolution[5] of a Cauchy function (or a number of these functions) with a Gaussian function (or a number of these functions), all with the same origin on the abscissa. Thereby the Voigt function allows an infinite number of profile-shape descriptions intermediate between pure Cauchy and pure Gaussian. This Langford paper stimulated a lot of further research adopting this type of profile-shape description. Ian Langford was and is generally seen as the one who introduced the Voigt function for profile-shape description. But see what follows.

To restore historical correctness and to demonstrate that recognition in science is not always lent in a fair way, a few remarks have to be made here. In the same issue of the Journal of Applied Crystallography, *in the paper immediately preceding the one by Langford*, the convolution of Cauchy and Gaussian functions to describe the shape of diffraction profiles was performed as well. This paper is authored by R.K. Nandi and S.P. Sen Gupta (*Journal of Applied Crystallography, 11 (1978), 6–9*). Admittedly the name "Voigt function" was not used in that paper. But in a number of earlier papers by Sen Gupta et al. the name "Voigt profile" was used already. As a matter of fact it appears that a convolution of Cauchy and Gaussian functions to describe diffraction profile shapes was proposed for the first time in the field of X-ray diffraction by F.R.L. Schoening in already 1965.[6] Langford did refer to the

---

[5] The "convolution", also called "folding", is a mathematical operation the explanation of which is beyond the scope of this booklet.

[6] Schoening did not use the name "Voigt function". He was apparently unaware (i) of its first use in spectral analysis in the field of astronomy by two Dutch astronomers already in 1947 (*H.C. van de Hulst and J.J.M. Reesinck, Astrophysics Journal, 106 (1947), 121–127*) and (ii) the work from 1912 (!) where this function was first considered by the namesake of this function (*W. Voigt, Sitzungsber. K.*

work by Schoening in passing, i.e. regarding a detail in his paper, not making clear at all that Schoening had already proposed the Voigt function so many years before him. Moreover he did not refer to the paper by Rama Rao and Anantharaman from 1965 (see Footnote 6) and also not at all to the work, published in a series of papers by Sen Gupta et al., all amply before 1978, where the Voigt function was applied in the profile analysis.

With a view to the above, the following observation is also remarkable. The manuscript of the Nandi-Sen Gupta paper was submitted on the 29th of June 1977. The manuscript of the Langford paper was submitted to the same journal, the Journal of Applied Crystallography, on the 28th of July 1977, so one month later than the Nandi-Sen Gupta manuscript. Strikingly both manuscripts were accepted for publication on the same day: the 16th of August 1977. At that time Ian Langford very often operated as referee for the journal. It is possible that he dealt with the Nandi-Sen Gupta manuscript as its referee, which, in light of the above, may give rise to second thoughts… I once asked Ian about this and he did not clearly deny it.

I have earlier indicated the occurrence of "inappropriate" referencing and, especially, the lack of referencing where required (see Footnote 3 in Chap. 6). Langford's paper has been and still is referenced a lot (more than 500 times since 1978; in the last years about 20 times each year). Thereby insufficient credit, if at all, is given to the true originator(s) of the use of the Voigt function in diffraction line profile analysis.

Line profile analysis requires (i) correction of instrumental broadening effects and (ii) separation of the size and mistake (strain) broadening effects. The more rigorous (but still based on more or less severe assumptions) methods were and still are cumbersome and require availability of at least two reflections (a basic one and a higher order one) of the same set of crystal-structure planes. Often more than one reflection is not available, as the higher order one is simply too weak or suffering from overlap with reflections from other sets of crystal-structure planes. Hence there is a strong need for a single-line (i.e. utilizing only one peak) method for the determination of size and mistake (strain) parameters. Against this background it was Staf de Keijser who initiated the development of such a practicable single-line method on the basis of only the breadth parameters of the measured profiles (so not the whole profile was analysed). The method assumes a Voigt profile shape and size and strain broadenings corresponding with Cauchy and Gaussian

Bayerische Akad. Wiss., 42 (1912), 603–620). At about the same time, but likely a few months later, the use of the Voigt function for describing the profile shape was also proposed by P. Rama Rao and T.R. Anantharaman (*Trans. Indian Inst. Metals, 18 (1965), 181–186*); in contrast with Schoening, they used the name "Voigt function" and explicitly referred to the paper by Voigt of 1912.

shapes. The three of us, Staf de Keijser, I and an enthusiastic student of mine, A.B.P. (Ton) Vogels, worked hard and eventually were successful and thus managed to develop an easy to apply procedure (a simple desk computer sufficed; nowadays a smart phone/pocket computer is more than adequate).

The next step was to write it up and submit the paper to a journal. We selected the *Journal of Applied Crystallography* (the obvious choice; see above). Our disappointment and annoyance was great when we got the paper back (from the editor Jerry B. Cohen) with a few remarks implying severe doubts regarding the usefulness of our method, although we had included a number of experimental examples showing that the method worked and was reasonably reliable. At this stage, in 1979/1980, we were starters in the field of line profile analysis and relatively unknown, which had, I still suspect, contributed significantly to the negative attitude of the original referee, who may well have been Jerry Cohen himself. This was the moment where we decided to contact Ian Langford (my personal acquaintance with him derived from the 10th IUCr (= International Union of Crystallography) Congress in 1975 in Amsterdam; see Chap. 5) and ask for his advice. We decided to extend the paper with a considerable analysis of possible error propagation. The paper certainly thus got a higher value, although the original results and their significance did not change at all. Langford had more experience in paper writing and that helped in reformulating certain passages. This is how Langford became co-author of our paper. We resubmitted the paper, and perhaps also because of the presence of Langford as co-author, the paper was now accepted straightaway (*Journal of Applied Crystallography, 15 (1982), 308–314*). It became a considerable success; the method was and is applied by many; the paper has been cited more than 1000 times, drawing, still, after 40 years, more than 30 citations per year.

Our interest in line profile analysis led through the years to further results. We developed a cooperation not only with Ian Langford (University of Birmingham) but also with Daniel Louër (University of Rennes); see Fig. 10.2. Together, "Delft" + "Birmingham" + "Rennes", we wrote a significant, joint chapter on line broadening analysis in the Rietveld method in the now classical book "*The Rietveld Method*" (edited by R.W. Young, Oxford University Press, 1993; see also Sect. 12.2).

The cooperation also led to friendly relationships. I was a couple of times in Birmingham visiting Ian and he came to Delft and stayed in my house where we spent some pleasant nights talking and drinking malt whiskey, that Ian preferred.

In an obituary on the occasion of Ian Langford's death in 2013 (at age 78), Professor Paolo Scardi (University of Trento, Italy) wrote about us as

Fig. 10.2 The "dream team" of line profile analysis. From left to right: Eric J. Mittemeijer, J.I. (Ian) Langford, Th. H. (Staf) de Keijser and Daniel Louër, at the *"Symposium on Accuracy in Powder Diffraction"*, organized by the National Bureau of Standards (NBS; now renamed as National Institute of Standards and Technology, NIST), in Gaithersburg, Maryland, USA (June 11–15, 1979) (*photograph provided by Paolo Scardi; photographer unknown*)

"*The "Dream Team" of Line Profile Analysis*", a designation that makes me somewhat melancholic as it awakens memories of a pleasant and fruitful time that lies now far behind me in the past.

Iron Nitrides and Iron Carbides; Tempering of Steels and Pre-Precipitation Processes

In order that a substance can be called a material it must have an importance beyond its mere existence. Hence, materials science is devoted to substances with a direct or indirect practical application. As an in this sense very outspoken case in point, and as a last example of the research performed by my group in Delft, I have chosen here the extensive research project dedicated to the precipitation processes of iron carbides and iron nitrides in supersaturated iron, which is of pronounced importance for the utilization of *steels*. The sketch of this project also allows the illumination of a few, more general experiences made by me.

Nitrogen and carbon are two elements which are neighbours in the Periodic System (see Figs. 5.1a and b). Their properties are, but only to a certain degree, rather similar. Thus they both dissolve interstitially in iron. This means that the dissolved nitrogen and carbon atoms occupy so-called *interstitial* sites in the iron crystal, i.e. between the crystal-structure sites where the iron atoms reside. This interstitial type of dissolution can be conceived as a consequence of the interstitial, nitrogen and carbon atoms being small as compared to the iron atoms. The iron–nitrogen and iron–carbon systems are of distinct importance in our daily life, in particular through the role of steels.

Steels are often used in a microstructural condition that can be crudely described as follows. The interstitial atoms are dissolved into the iron at elevated temperature. Upon rapid cooling (called quenching), say to room temperature, the iron atoms want to adopt another crystal structure, which desire is obstructed by the present interstitial atoms. As a result a non-equilibrium state develops, called *martensite*, with the iron atoms arranged according to a so-called body centered tetragonal crystal structure and the interstitial atoms occupying a number of sites of only one type of the three types of octahedral interstitial sites in this crystal structure. On this one type of octahedral interstitial sites the interstitial atoms are randomly distributed, i.e. no regular, periodic arrangement of the interstitial atoms occurs.

The reader with no background in materials science and crystallography needs not to comprehend fully the contents of the above paragraph. The crucial recognition is that the martensitic structure is highly unstable, i.e. states of lower energy are strived for by the material. Such energy lowering is brought about by structural rearrangements. It depends on, say, the temperature, if the necessary atomic mobilities can be established. The materials scientist says: the kinetics of the process (of atomic rearrangement) determines if the process can occur significantly at the conditions (as the temperature) pertaining to the system.

The martensite is very hard and, as is well known for hard materials, also rather brittle, which obstructs practical application. As a remedy the martensite is annealed giving rise to a series of consecutive precipitation (of iron carbides or iron nitrides) processes. The process is called "tempering". It results, by tuned application, in more ductility accompanied with an only modest loss of hardness. This resulting microstructure has led to the very successful commercial exploitation of steels.

Whereas the iron–carbon system had been the focus of a large amount of work in the literature (reflecting that most steels contain, next to iron, carbon as alloying element), not much work on the iron–nitrogen system had been performed, but its importance gained more and more importance as a consequence of the widening application of nitriding processes (see Chap. 8) and the rise of the so-called "high nitrogen steels".

At room temperature already, iron–carbon and iron–nitrogen martensite show that the interstitial atoms have mobility enough to establish structural changes; the iron atoms stay at their crystal-structure sites at this temperature. These room temperature processes in iron-based martensite are highly intriguing and have fascinated materials scientists already many decades, leading to a bunch of controversial and opposite interpretations. This topic of possible interstitial rearrangement also captivated my mind. Therefore, after being introduced to this topic, during my time in Korevaar's group, by my Polish colleague Jacek Wierszyllowski and our initial work on iron–nitrogen martensite (see Chap. 8), I decided to develop a research program on the processes occurring in iron–carbon and iron–nitrogen martensite at room temperature and upon annealing.

In this project I had my first experience with a Chinese student. Liu Cheng first did his Master's under my guidance and then continued with his Ph.D. in my group. He was a remarkable young man; he was talented with an almost incredible devotion to his research. He was initially rather shy, couldn't well speak and write English. He treated me with a courtesy I was unused to: he stood up when I entered his office.

In my group we were rather informal: nobody addressed me saying "Professor". However, with the exception of one technician, who said he wanted to keep distance. I understood that attitude as preserving for him the possibility for being able to easier criticize me, if felt necessary, which, however, did not happen.

It took some time before Liu Cheng had adapted himself to our informality, but he remained in an outspoken way polite towards me. In this respect, later experiences I had with Chinese students and post-docs in Stuttgart were not much different from my first encounters with Liu Cheng.

Evidently, the awe for Professors in China appears to be much larger than in (West-)Europe. The same seems to hold regarding the respect for elder people, but this could not apply in case of my relationship with Liu Cheng, as I was only a little more than 10 years older.

The quantity and quality of the work done in a little more than 4 years of research and as expressed in a series of papers by Liu Cheng et al. is impressive. His new results in fact provide an original (and comparative) overview of the tempering processes in iron–carbon, iron–nitrogen, iron–carbon–nitrogen and iron–nickel–nitrogen martensites.

Here, I will focus on the results regarding the *pre-precipitation* processes in the martensites, with an emphasis on a still existing controversy prevailing in the literature.

Liu Cheng et al. succeeded in demonstrating that a number of carbon-redistribution processes occur at room temperature in iron–carbon martensite. The dominant part of the carbon atoms is involved in the development of local enrichments of carbon atoms. These clusters are conceived as aggregates of iron and carbon atoms containing different amounts of carbon atoms distributed randomly over the one type (see above) of octahedral interstitial sites. This absence of long range ordering of the carbon atoms implies that the carbon enrichments must be called carbon *clusters*. In case of the iron–nitrogen martensite, the dominant part of the nitrogen atoms is, as holds for the iron–carbon martensite, involved in the development of local enrichments of nitrogen atoms. However, in striking difference with the iron–carbon martensite, these nitrogen enrichments do exhibit a long range ordering of the nitrogen atoms on the one type of octahedral interstitial sites, an ordering as compatible with α'' iron nitride, α''-$Fe_{16}N_2$ (see also the discussion in Sect. 9.3). Therefore these nitrogen enrichments are nothing else than α'' iron nitride *precipitates* (*Metallurgical Transactions A, 21A (1990) 2857–2867, Metallurgical Transactions A, 22A (1991) 1957–1967*).

It has been suggested in the fifties by Ken Jack (cf. Sect. 9.3) that the first precipitate to develop in iron–carbon martensite must be α'' iron carbide, α''-$Fe_{16}C_2$, i.e. an iron carbide fully isostructural with α'' iron nitride, α''-$Fe_{16}N_2$. Apparently, the parallels of the iron–carbon and iron–nitrogen systems have such an appeal that later researchers have more or less indiscriminately adopted this suggestion as representing reality. However, nature is not always such simple.

To prove the *non*occurrence of α'' iron carbide, α''-$Fe_{16}C_2$, and the occurrence of α'' iron nitride, α''-$Fe_{16}N_2$, in the next Ph.D. project, of M.J. (Marc) van Genderen, synchrotron radiation at Daresbury (United Kingdom) was applied. The much higher X-ray intensity available there, than applicable at the Laboratory with a conventional diffractometer, in principle allowed the detection of the superstructure reflections of α'' iron nitride, which are solely determined by the ordered arrangement of the nitrogen atoms and which were too weak for detection in the Laboratory. Thus the occurrence of α'' iron nitride precipitates, and not nitrogen clusters, in the initial stage of aging of iron–nitrogen martensites, was definitively proven. It was also demonstrated that the carbon enrichments, occurring in the initial stage of aging iron–carbon martensites, did not show the ordering as for the nitrogen atoms in α'' iron nitride. This work was followed by high resolution transmission electron microscopy confirming the above and demonstrating moreover that the developing carbon enrichments, i.e. the carbon clusters, were (indeed; see above) of variable composition; the development of carbon-enriched and carbon-depleted zones eventually led to the occurrence of a periodic modulation in the material (*Metallurgical Transactions A, 24A (1993) 1965–1973, Philosophical Magazine A **81** (2001) 741–757*).

One may wonder, and we did, why the iron–carbon and iron–nitrogen systems are such different here. This may be due to the difference in elastic strain energy released upon potential ordering of the interstitials in an interstitial enrichment: the amount of such energy released would be larger for the case of the nitrogen enrichments.[7]

This tiny detail, the intriguing difference of the behaviours of carbon and nitrogen in iron, that has fascinated me, is of course not a groundbreaking result in the course of science. I have recounted this story here also because it allows me to describe, by an extreme example, the sometimes very sloppy way present-day scientists consult the preceding literature and thereby ignore the results already obtained by their predecessors (also see Footnote 3 in Chap. 6):

On the basis of the above, one would expect that, after 1993, in papers, dealing with decomposition of iron–carbon martensite, the α'' iron carbide, α''-$Fe_{16}C_2$, does not appear. Unfortunately this is not the case. In many cases it was groundlessly assumed that α'' iron carbide, α''-$Fe_{16}C_2$, developed upon decomposition of Fe–C martensite. The "evidence" presented to this end can be shown to be flawed or it represents merely "wishful thinking". Apparently the parallel of iron–carbon and iron–nitrogen is too seductive... Yet, it is amazing, the more when the results by Liu Cheng et al. and van Genderen et al. are not even referred to and, if deemed necessary, possibly criticized. A very striking case occurred very recently:

---

[7] Carbon in iron–carbon can have some positive ionicity due to transfer of electrons to the surrounding iron atoms, whereas nitrogen in iron–nitrogen can remain neutral. Consequently the "size" of the nitrogen atom in iron–nitrogen can be a little larger than that of carbon in iron–carbon. The release of elastic strain energy upon ordering of the nitrogen atoms according to the α'' crystal structure would then be larger than for the carbon atoms upon (hypothetical) similar ordering.

In 2019 a paper was published, on decomposition of iron–carbon, in the same journal where the Liu Cheng et al. and Marc van Genderen et al. results were published. The authors did describe, without any proof, the developing carbon enrichments as α'' iron carbide, α''-$Fe_{16}C_2$. Moreover they did not cite at all the previous Cheng-van Genderen et al. work published in the same journal, which, if they had done that, of course had to imply either to adopt the conclusion that α'' iron carbide, α''-$Fe_{16}C_2$, does not occur or to criticize the Cheng-van Genderen et al. results. In view of the availability of electronic literature surveillance systems and certainly regarding preceding work published in the *same* journal, this is an extreme example of lack of referencing.

In this case it was even more painful as the authors were materials scientists at the university where, about 20–30 years earlier, the preceding work had been performed. Moreover, the senior author of the 2019 paper and I did know each other for a similarly long time. For this one time in my career, I decided to respond to this case where work by my co-workers and myself was unjustly and painfully ignored. I wrote a friendly and carefully formulated email to my colleague, which started with making clear that I had written this email only after having overcome great hesitation. I also did not require a correction in the literature and did not plan to submit a "comment" to the journal concerned. Until the day of writing this chapter I have not received any response.

A visit to the Bergakademie Freiberg; die "Wende"; a Dark Corner in Science

In probably the first or second year after the "Wende" of 1989 (= "the Turn of 1989"; i.e. the peaceful revolution leading to the (re)unification of the "Bundesrepublik Deutschland (BDR)" and the "Deutsche Demokratische Republik (DDR)"), I travelled to Freiberg in the former DDR for a visit to the Bergakademie Freiberg (= mining academy). I was invited by two colleagues: Professor Heinrich Oettel and Professor Peter Klimanek.

I was familiar with Heinrich since we met the first time in 1976 at the VIII Hungarian Diffraction Conference in Tihany. At that time I still was a Ph.D. student. On that occasion we had had a first intensive discussion on the merits of the Warren–Averbach method (which, at the time, was at the focus of my attention in Delft) and the Krivoglaz-Wilkens approach (applied by Oettel and Klimanek) for the diffraction analysis of crystallite size and mistakes in solids. My memory here is especially sharp as in that discussion (that, remarkably, took place during the conference excursion: a boat trip on the Balaton lake (= "Plattensee")) also Professor Rolf Hosemann (Director at the Max Planck Institute (Fritz Haber Institute) in Berlin) participated, who was well known for his theory on paracrystalline order and its diffraction analysis.[8] Such a discussion, for a young scientist, as I was, leaves a lasting impression.

I had been before in Freiberg, before the "Wende". Not only the diffraction analysis group of Oettel, the more practical man, and Klimanek, the more theoretical man, had attracted me, but also the comprehensive nitriding research led by Professor Spies had my strong interest as well. Although the Bergakademie Freiberg had a privileged position in the DDR, during my first visit I could not avoid to notice that in no way the facilities could compete with those available to me in Delft.

My contacts and cooperation with scientists from behind the Iron Curtain has made me aware of the luck I had to perform my science in West-Europe. It is not often said aloud, but, back then, one could not avoid the general impression that the non-availability of top level equipment and thus the impossibility to attain high/highest precision and accuracy in experimental data had led to sloppiness in attitude in a significant amount of the experimental scientific work performed behind the Iron Curtain. At least that is what I observed upon reading and studying quite a number of papers from that part of the world in these days.

Now back in Freiberg, after the "Wende", the difference in the atmosphere with my previous visits was enormous. Already my hotel now approached "Western" standards... The first day Heinrich and I made a walk through the small, beautiful, old city centre. On a square Heinrich made a stop at a building that appeared to be the town hall. It was a warm day and the doors had been opened. Heinrich said: "Eric, look!" I looked through the open door he pointed at. I saw a group of people apparently having a meeting. In their middle sat, to my great surprise, a man I knew: Peter Klimanek! I was told that after the "Wende" the political engagement of Klimanek had prompted him to be a candidate for the town council, apparently with success.

---

[8] R. Hosemann and S.N. Bagghi had authored a well-known book: *Direct Analysis of Diffraction by Matter* (North-Holland Publishing Company, Amsterdam 1962). The book had gotten a remarkable, not really praising, review by P.M. (Pim) de Wolff (a famous name in the fields of X-ray powder diffraction and incommensurately modulated atomic structures (see Footnote 2 in Sect. 9.1 and see my book *"Fundamentals of Materials Science"*, Second Edition, Springer, 2021; Sect. 4.8.1)). De Wolff, Professor at the Delft University of Technology, was also the Chairman of the committee that nominated me for the Burgers Chair. De Wolff wrote, referring to the symbols used by Hosemann and Bacchi: „Jedenfalls geben sie dem Leser oft das Gefühl, eine zentralafrikanische Sprache zu studieren, anstatt Physik." (= In any case they give the reader often the feeling to study a language of Central Africa, instead of physics.).... (*Berichte der Bunsengesellschaft für physikalische Chemie, 67* (1963), 134–135).

As a matter of fact, and in line with the above, Klimanek had maintained during the DDR reign a "political" position more or less openly independent of the "party line" (he was no SED ("Sozialistische Einheitspartei Deutschland) member), which had obstructed him also in getting a Chair. This was now "corrected" after the "Wende": in 1992 Klimanek was appointed Professor. Shortly before Oettel had been elected as the Director of the newly founded Institute for Metallurgy and was appointed Professor in also 1992. Former department heads, somehow linked with the SED regime, had been removed from these positions. In these days such transfers occurred at many places in the former DDR.

The level of science performed in the DDR was at many places of mediocre quality, especially in the humanities and social sciences. Also in the natural sciences high, international level was an not the rule. However, exceptions did occur. One of these, of which I am aware, was personified by Joachim Kunze, a scientist of relatively low formal rank and of modest personality in the "Zentralinstitut fuer Festkörperforschung und Werkstoffwissenschaft" in Dresden, a part of the (after the "Wende" dissolved) "Akademie der Wissenschaften" of the DDR. He had performed largely theoretical work on the thermodynamics of the iron–carbon–nitrogen (Fe–C–N) system. He worked more or less in "splendid isolation" and published his work as a monograph: "*J. Kunze: Nitrogen and Carbon in Iron and Steel. Thermodynamics, Physical Research Vol. 16, Akademie-Verlag, Berlin, 1990*". That publication is not well known (e.g. the Web of Science does not retrieve it).[9] After the "Wende" he published some main and updated results of this comprehensive work also as a paper in German in a German journal (*HTM Härterei-Technische Mitteilungen, 51 (1996), 1996, 348–354*), which was the result of the presentation he gave, close to retirement age, at a conference on nitriding and nitrocarburzing (24th–26th April, 1996, Weimar, Germany) that I, together with Johann Grosch (Technical University of Berlin), had organized. Both publications did not promote wide dissemination of his results. But a few experts noticed it.

Kunze serves as an example of those deserved scientists, in this case from behind the Iron Curtain, who as scientists are unjustly overlooked, as their

---

[9] Books published by the Akademie-Verlag, Berlin, in the DDR, mostly dealt with the humanities and social sciences. These books were not very well incorporated in the BRD (West-Germany) owing to obvious "political" differences as manifested especially by those publications touching the history and consequences of *divided* Germany. Further, after 1990 the downturn of the publisher started by the first of a number of consecutive sales to Western publishing houses. In 2013 the Akademie-Verlag ceased to exist. These aspects and this development did not support promotion of Kunze's book that moreover appeared in the for the Akademie-Verlag turbulent year 1990.

results are.[10] I learned of Kunze's work in the beginning of the eighties at the conference in Poland that I already mentioned in Chap. 8. Much later, and then in Stuttgart, the contact with Kunze (then already retired) was renewed, as we had embarked on a larger project of our own that was concluded, practically at the end of my career, with a full model of the thermodynamics of the iron–carbon–nitrogen (Fe–C–N) system (*Metallurgical and Materials Transactions A, 47A (2016), 6173–6186*). The work by Kunze had served as a reference point, which I want to acknowledge, not only as realized in our paper, but gratefully also here.

The task of a Professor; Evaluations and final years in Delft

I had decided from the start that I would not become a Professor ending up as a more or less full time administrator, of which I had seen examples. I wanted to intensively guide and cooperate in the scientific work with my students and co-workers. This obviously is a time consuming interpretation of the supervising task of a Professor. Most of my days were taken by meetings and discussions dealing with the science produced by my group. I have never ceased to love this dominating part of my life. It also meant that even more than before I spent evenings and part of the weekends almost invariably with reading reports, modifying and writing manuscripts and reading literature.

Lecturing for students was a task that I enthusiastically performed as well. If possible, own research experiences were woven in. Now and then I was confronted with a question by a student that allowed me, by thinking it over, to understand and present better than before a point dealt with. Also Professors can learn by teaching…; students highly profit from such encounters.

---

[10] It is often said that, even if a result of importance has been published in a remote, dark corner of the building of scientific literature, that even then sooner or later it will be found. That may be true, but I would not be too sure about that, in view of the present-day rather wide-spread unwillingness to check comprehensively the preceding literature on a topic (cf. Chap. 6). In any case, it can be too late for the originators of the finding concerned, as in the time passed the same result can possibly have been obtained, independently, by someone else who earns all the credits.

Kunze has drawn a marginal number of citations. Yet he has been a scientist of significance. The fate of nonvisibility may be shared by those scientists who with great perseverance, perhaps almost stubbornly, have stayed with a topic that by the mainstream of scientists is not considered to be rather "hot", "modern" or "popular" (to a certain extent this can be said from Kunze and his work) and accordingly had to publish their work in less popular, lower quality journals (Journals as Nature and Science, and others, do consider also the degree of *instantaneous* impact of a paper as a criterion for publication, which can lead to the refusal to even let a very good piece of science to be refereed…). We should not forget that the diligence and zeal of such scientists to stay with their "unpopular" topic in some cases can lead to novel work that "out of the blue", unpredictably produces extraordinary results of lasting impact on science.

From this time I have the strong conviction that active researchers are the preferred teachers.

Of course, a number of administrative tasks could not be gone out of the way, both within the Laboratory and within the Faculty (at this time the Laboratory of Materials Science had become a part of the now joined Faculty of Chemical Technology and Materials Science), and also within the scientific community at large. I generally did not at all strive for such tasks...

In the year 1996 I and my colleague Laurens Katgerman, both from the Laboratory of Materials Science, had been appointed as the Board of the Faculty. This was remarkable as no Professor from the much larger chemistry part of the Faculty would be a member of the Faculty Board. But so it happened: Laurens became Dean and I came to be Vice-Dean. I was to remain in this function only until my leave to Stuttgart at the end of 1997, but that was unknown at the time of appointment.

Laurens and I knew each other well; we had published together on rapidly solidified metals during the time I was in Korevaar's group. Upon his appointment as Professor and return to Delft in 1992, after his stay with industry abroad from 1983 till 1992, we renewed our contacts. Laurens was an ardent violin player. I was an enthusiastic piano player. We decided to make music together and from that time on we regularly spent evenings in my home playing violin-piano sonatas, as, particularly, those by Schubert and Dvořák. It was a very enjoyable experience. We also liked to drink a glass of wine and spoke a lot, unavoidably also about our experiences at the Laboratory and in the Faculty. I keep most pleasant memories of these evenings. Later in Stuttgart such a relationship with a colleague did not develop.

In order to develop and maintain a successful group at a university the scientist in charge and his/her co-workers must spend a considerable amount of their time on drafting research proposals to be submitted to funding, government-related or European Community agencies. In such proposals one applies for personnel positions (usually Ph.D. student and post-doc positions), equipment and consumables. The success rate of such proposals is usually very low, nowadays (even distinctly) less than 30%.

It was always a nerves loading activity to arrange (in time) for the necessary positions, equipment and consumables. It absorbed a lot of my energy that thereby was not available to scientific work.

A university should have the possibility to provide any Professor with, say, a reasonable number of Ph.D. students and corresponding finances. In quite a number of countries, in any case in The Netherlands and Germany, proven, also repeatedly confirmed quality of a researcher at a university is not rewarded such that a certain amount of funds is put at his/her disposal on a continuous basis: each time, each year he/she has to write and submit proposals, again and again, already only to allow the bare survival of his/her research. I find that ridiculous. Too much time and energy get lost in this proposal fight for means by too highly qualified scientists. This annoyance eventually led to my move from Delft to Stuttgart.

My personal pinnacle of annoying experiences, associated with acquisition of external funds, is presented by the following tale.

At the end of the nineties I had to go, late summer, to the south of Europe for a "midterm review", i.e. an evaluation halfway, of a European research project, in this case incorporating my group as participating partner. The meeting took place in the presence of an officer of the European Community who came from Brussels. His task was to assess our performance. His judgement would be of utmost importance in order to get access to the remaining part of the originally assigned funds. Hence, I, and the other project leaders, had to attend and present our work.

The night before the actual meeting we enjoyed a delicious dinner in a fish restaurant, at the cost of the project, worked on the social contacts and became acquainted with the "officer from Brussels", who was unknown to me. Unsolicited the man told me that he was an expert in our field—perhaps because doubt was possible (indeed his name did not ring a bell with me)—and that he always travelled business class…

The next day the project participants, about 20 people, and the "officer from Brussels" gathered in a large conference room of the local university. I was the first to lecture on our work. Shortly after I had begun to speak, the "officer from Brussels" started a discussion with his neighbour. Especially upon this discussion becoming louder and louder, it drew the attention of the entire audience. I could no longer bear it. I stopped with speaking, as my lecture, after all, in particular was intended to inform the "officer from Brussels", and looked vigorously in the direction of the "officer from Brussels". He did not notice it and, to great amusement of all attendants, continued his discussion. Then he suddenly noticed that something was wrong: the background noise, that I had been taking care of, had fallen silent… He looked up. I said: "Have you finished? Thank you, then I can continue." Without further "incidents" I managed to arrive at the end of my talk. Believe it or not: the next speaker had embarked on his presentation for only 10 min, then the "officer from Brussels" fell asleep. After the conclusion of the third and last lecture of the morning session he woke up: it was time for the lunch. Luckily, in the end we did get our money.

Recalling this admittedly extreme event still gives me the feeling of being prostituted on the spot. Very little is contributed to science by these meetings, which moreover cost a lot of travel and accommodation funds (all incorporated in the project budget plan from the beginning of the project...). Every effort is done to get a positive evaluation (I well remember soft criticism by a colleague regarding my "action" towards the "officer from Brussels", that, he thought, might have endangered a positive outcome of the evaluation (it didn't)). In any case the dependence on a positive review drives the scientists involved to unnecessary meetings and delicate handling of officials from funding agencies, not to speak from the associated enormous administrative burden for a running, especially European Union project. This all distracts from the genuine scientific assignment.

During the first half of my career I have witnessed that evaluations of research performance have become ever more important. To get rid of or sort out "fallen asleep" scientists and "dormant", non-productive groups, which one here and there could come across decades ago (in the sixties, seventies and eighties, as I have witnessed myself), this instrument of evaluation can be effective. Application of evaluations becomes more problematic if distinction between good to very good scientists has to be made. This holds the more realizing that evaluations in science are made, of course anonymously, by the scientists themselves: the proposal drafters judge the proposals drafted by their colleagues. This is what is called in science the peer system.

The same system is applied to manuscripts submitted to a journal: scientists, active in the same field as that covered by the paper, evaluate the quality of the manuscript and advise the editor of the journal to either accept or let revise or reject the manuscript.

In scientific communities sufficiently small so that the scientists know each other personally, such judgements by the peer system can be more or less biased by these personal relationships: it is easier to formulate a negative opinion about a proposal or manuscript from a "fiend" than from a "friend". Also envy and prejudice in general can lead to unfair evaluation. There are more objections possible with respect to the peer system. Especially if minor quality differences become important, as between a group of good to very good scientists and their proposals, such biasing parameters can become of decisive importance. This is most regrettable. However, no viable, acceptable alternative has been found; we have no better, more honest instrument than judgement by peers. Yet, for example, an editor of a journal must have a large sensitive spot for biased referee reports (cf. the case discussed in Sect. 9.3).

My group in Delft was part of a community of materials science groups within the Foundation for Fundamental Research of Matter (FOM). FOM was a funding agency of essential importance for these groups (my own Ph.D. position in 1972 was applied for by my doctor father Okkerse and granted by this community within FOM). Each year or every other year these groups could submit proposals for support for specific projects. The growing "need" for evaluations once led this community of modest size to induce an overall evaluation of the groups by especially high quality, "external" examiners, where "external" here also means: from abroad. In this way a conceivable objection of self-evaluation was remedied. Thus Professor Frans Spaepen was asked to act as an external examiner. Spaepen is Professor at the elite Harvard University (Cambridge, Massachusetts, USA) and from Belgian, more precise Flemish, origin. The latter aspect had the advantage one could speak Dutch as well. Thus Frans, accompanied by me, travelled from group to group. These were tiring days, but also very informative and interesting. I noticed how attitudes and discussions were influenced by the presence of Spaepen; too much depended on a "good outcome" for the group visited. This is of course fully understandable, but being this time "on the other side of the table" made an enduring impression on me. After such a day we were exhausted. I remember well how Frans and I relaxed, in the warm evenings, on a terrace in front of a bar, drinking one or two Belgian beers, where Frans, naturally, was an expert.

The still growing "need" and wish from funding agencies and university boards to more and more evaluations of research performances, eventually led in 1995 and 1996 to a nation-wide evaluation of all chemistry and materials science groups. The organizer was the Association of Universities in the Netherlands (VSNU = Vereniging van Samenwerkende Nederlandse Universiteiten). Also in this case external examiners were the judges. This evaluation cost an enormous amount of time. A very large report had to be submitted wherein the research performed by the whole group and its outcome, also in terms of papers etc., had to be presented. It was the first time that I, and my co-workers, had to produce such a voluminous report about our science. Of course the importance, in view of possible consequences, was correctly seen as enormous. In December 1996 the "verdict", i.e. the detailed judgement, would be sent to me. One evening I sat late in my office going through the post of the day, as I had had no time earlier to do that. Upon opening a thick envelop I realized what its contents would be: the summarizing, final judgement immediately struck my eye: we had obtained the highest possible marks. I still feel the sensation of enormous satisfaction I experienced at that moment. Of course I did realize the only relative significance of the opinion of a few judges, but I knew this was very good for the group and for me.

It turned out that at the university only one other group had achieved such similarly positive evaluation. The result was publicized nationally and several people sent me notes with congratulations. My doctor father, Professor Okkerse, sent a letter that indicated how pleased he was with this result for the group he once had led himself more than 20 years earlier.

The reader by now should have the indeed correct impression that my time on Burgers Chair generally was a good, happy and successful one. Unfortunately, one sad experience must be recalled too.

My contact with Kees Brakman (see Chap. 5) had become more intensive after I had left Korevaar's group, but our contact now took place on a more professional level as well: Kees became my successor in Korevaar's group and, from that time on, he and I cooperated on a number of projects, thereby strengthening the bond between my group and that of Korevaar, a connection that I had already maintained, from the other side, during my time in Korevaar's group.

Kees and I thus correctly deduced the intrinsic kinetic parameters for the case of coupled growth of two different compound layers on top of a substrate: here iron–boride layers on iron-based substrates upon boriding, whereas in the boriding literature until then the consequences of the growth coupling had been overlooked, which had grave consequences for the values given for the kinetic parameters. Our finest piece of cooperation involved the internal nitriding of iron–aluminium alloys. This was the Ph.D. project of Mohammad Biglari (also see Chap. 13).

Kees was tall and broad (distinctly taller and broader than I am, and I am not small) and muscular. I have always seen a relationship between his appearance and his activity as a reserve-officer in the Dutch army. To become and remain a reserve-officer regular training stages were imperative. Thus, once per year, or so, Kees had to go "on camp". Each time he returned it appeared he had obtained a higher military rank. He ended his career as reserve-officer with the for a reserve-officer high rank of "major". Indeed, he had also some military traits in his performance as a civilian as well. Against the sketched background his hiking along less used, demanding tracks through the mountains of Europe can be considered. He did this alone. In August 1994 he did not return to the Laboratory from one of his long hiking trips at the day we had expected him. At first we were not very worried, as we knew Kees was an experienced hiker in the mountains. But soon we started to find out what could have happened. Kees had disappeared during a trip in the Pyrenees. The last mountain cabin where he had registered was known; he had not arrived at a next one. Somebody had heard someone yelling for help, on a few hours walking distance from the last mountain cabin that had recorded his visit, at the day Kees had left this mountain cabin. Extensive searches were made but they were in vain. I was informed that eventually the Dutch airforce sent a reconnaissance jet making high resolution photographs of the region; supposedly that was only possible because Kees was a high ranking reserve-officer. Kees could not be detected. The whole Laboratory followed the process; everybody knew Kees. The winter came and further search actions became impossible. In the following spring of 1995, at last, what was left of Kees was found in a gorge... A moving memorial service was organized at the university. Mohammad was deeply touched. A few months earlier, after the disappearance of Kees, he had defended his Ph.D. thesis publicly and also on that occasion he had referred in an emotive way to Kees.

I was now 10 years on Burgers Chair, 25 years active at the Laboratory, 45 years old and asked myself if this was the position to stay at until my retirement some 20 years later. Some text parts in the above may already have identified some developments that I considered less advantageous. In that context the following is remarked.

My group was relatively large, requiring rather extended efforts of fund acquisition. A number of years before the time indicated above, we wished to replace our aging transmission electron microscope by the newest high resolution transmission electron microscope (HRTEM). We wrote an application. The corresponding investment was that large, that, as response, FOM decided that we should reformulate our application such that we would ask for founding a *National Centre of High Resolution Electron Microscopy*, as part of my group. This National Centre then had to deliver service research for all researchers in The Netherlands desiring high resolution electron microscopical investigation of their specimens. This was not what I wished: I only wanted a new HRTEM for my group..., but I had no choice. So we wrote a new, still more voluminous, application and made a tour along groups in The Netherlands in order to get their support. In the end the application was granted, which meant that instead of one we could buy two new microscopes and we also got additional personnel positions.

Whereas others in my position may have enjoyed being the boss of now also a "National Centre", I saw problems emerging at the horizon. I anticipated that in due course a certain degree of disintegration of Burgers Chair could occur, especially if this "National Centre of High Resolution Electron Microscopy" would be seen by the outside world as an independent entity, apart from the already described tendency in the faculty to look first at the largest group if personnel and finance savings had to be realized. Moreover, the increased responsibility for now also the *National Centre of High Resolution Electron Microscopy*, that needed costly investments also in forthcoming years, required an even more increased activity in writing performance reports and applications for support. Of course our own research also profited strongly from the application of the new microscopes in our research and we were successful in evaluations, as I described above. However, a process had started in my brain that made me sensitive for other jobs where I would be free from comprehensive fund raising assignments.

At the same time I had decided for myself that I wanted to stay immersed in science, i.e. I did not want to develop myself in an administrative direction. I also believed that I needed a drastic change, especially of environment; to stay another 20 years at the same position in the same Laboratory leading the same group might cause some degree of "arteriosclerosis" and insensitivity for new impulses.

At this point I became aware of an open Director position at the Max Planck Institute for Metals Research in Stuttgart (Germany). The Max Planck Society selects scientists of worldwide top level for their Director positions at their institutes. These scientists then are provided with a personnel and financial budget of permanent character that makes it possible that they can perform their research with a substantial team, in complete freedom regarding what research to perform, until retirement. There is no other research organization in the world, that I am aware of, that offers such support, on comparable level, to eminent scientists. This is rather different from the eternal fight for funds I had experienced until then and what irritated and distracted so much. I did and still see no reason why scientists of repeatedly proven high quality have to be subjected to the perpetual fight for means. The objection to this thought, that is normally brought forward, is that guaranteed, lasting support would make scientists lazy, less productive and less creative. The Directors of the Max Planck Society by their performance show that that is a bogus argument.

The development of my contacts with Stuttgart developed slowly and rather secretly; only my secretary and very, very few others knew. Only after, at last, the Max Planck Society had formally offered me the Director position in Stuttgart, implying that within the Max Planck Society this information was "public", spreading of this news in the materials science community became unavoidable. I simply waited to be addressed about that, as I had started negotiations about the conditions for my move, so that a definitive decision had still to be made from at least my side.

Then my father suddenly and unexpectedly died (see Chap. 1). A short time before, without realizing that it would be the last time that we could speak with each other, we had enjoyed a dinner at the fish restaurant in Ijmuiden at the North Sea coast, close to the sluices at the end of the North Sea Canal, not far from Haarlem (see Chap. 3), a favourite dining place of my parents. I had informed my parents about my likely move to Germany and had explained to them my considerations leading to this possible decision. Understandably, they could not well appreciate what it meant to me, notwithstanding that they were proud of me.

During this time the university (TUD), knowing of the offer made by the Max Planck Society, made an effort to keep me in Delft. This offer meant a salary increase, even somewhat above that presented by Stuttgart, and a guarantee of an appreciable amount of means, but, and that would finally be decisive, for only five years. The university board may not have fully understood my motivations for moving to Stuttgart: it was primarily the guarantee of permanent and amply support of personal and finances; my personal salary did interest me only secondarily as long as it would not be less than the current one.

These events delayed my decision, but in the end I decided definitively to accept the position in Stuttgart.

I am not fully sure about the reaction and feelings of my group once they understood I would leave them. Nobody expressed any criticism against me, on the contrary. Obviously, they realized, quite correctly, that for them a large degree of uncertainty and change was associated with the near future. But that the Laboratory of Materials Science, as a unit, would cease to exist and literally demolished some nine years later, in 2006 (see Chap. 8), was certainly beyond anybody's imagination. On the occasion of my move, my group offered me an atomic structure model, with inscription, of α'' iron nitride (α''-$Fe_{16}N_2$), a compound that had played a central role in my research especially during the last years in Delft. The model still stands on my writing desk in Stuttgart, reminding me of many good and successful years in Delft.

So, at the end of 1997 and formally at the start of 1998 I began in Stuttgart a new phase in my life.

Upon leaving the Delft University of Technology, after 12 years as Professor, 25 dissertations had been concluded under my supervision. Apart from the topics touched upon before, as diffraction analysis of mistakes in solids, the iron–nitrogen interaction and (pre-)precipitation processes, we focussed also on the role of mechanical stress and reactions in thin films and the initial stages of oxidation of metals. In retrospect these last topics, i.e. first steps in "nanomaterials" and interface science, can be considered as a prelude of much of the work performed later by my department in Stuttgart.

Three of my "Delft" Ph.D. students later became Professor in their own right:

– Marcel Somers, whose Ph.D. project was devoted to nitriding, more or less stayed in this field and at the Technical University of Denmark (Lyngby) kept contact with me (then in Stuttgart) which later resulted in a jointly edited book on "*Thermochemical Surface Engineering of Steels*" (Elsevier-Woodhead Publishing, 2015);

- Bart Kooi, whose Ph.D. project was dedicated to the thermodynamics of iron–nitrogen phases, now is Professor at the University of Groningen; and finally,
- Rinze Benedictus, whose Ph.D. project was devoted to solid state amorphization, now is Professor at the Delft University of Technology.

The above text illustrates that not only the research results are outcome of the work performed by a scientist at an academic institution: the education of young people in their formative years is an important task as well. Then looking back on these "Delft" years, I may be forgiven that some satisfaction in also this sense is felt by me.

11

Tarnished Science

Abstract "Clean" science versus "tarnished" science. Three types of tarnished science are identified: (i) bending of the results of research to serve the interests of the employer or sponsor, (ii) the financial support is from a detestable source, and (iii) immoral behaviour of the scientist. Exemplary, corresponding examples are provided by clinical research projects supported by the pharmaceutical industry, the financial support by Jeffry Epstein to elite universities in the USA and the professional life and works of the "father of the atomic bomb" in Pakistan.

Scientific research ideally is a clean activity. This noble aspiration requires that (i) its course is not influenced by parameters other than the curiosity and decisions of the scientists themselves, who, moreover, (ii) do not strive for personal gain by serving interests beyond the progress of science. This picture, of course, is untenable as indicative of present-day reality.

A scientist as Charles Janet (see under "*Looking at The Periodic System; The Role of "Outsiders" in Science*" in Chap. 5) was a "man of means", with both enough time and finances to perform successfully scientific research as a private person. He chose his topics for research completely independently; he simply did exclusively as he liked. Nowadays scientists are only rather rarely or not at all in such a position. It starts with the influence exerted by their employers.

Obviously a scientist working for a firm, usually with commercial interests, has to comply with a field of research as demarcated by his employer. Within this "fence" he/she may have more or less freedom to move. If the employer is the government, e.g. in the form of a university or an institute,

E. J. Mittemeijer, *How Science Runs*,
https://doi.org/10.1007/978-3-030-90095-3_11

the independence of the scientist is relatively large, but often not limitless; the "fence" is far away but observable.

The funds at a university, as provided by the government generally and worldwide, do *not* allow for sufficient positions (if at all), as for Ph.D. students and post-docs, and for (expensive) equipment. The scientist has to apply for that money, e.g. within the framework of self-governing governmental programs, as those of the Deutsche Forschungsgemeinschaft (= German Research Society (DFG)) or of the National Science Foundation (NSF) in the USA, or of supranational programs, as founded and financed by the European Union (EU) (also see the discussion in "*The task of a Professor; Evaluations and final years in Delft*" in Chap. 10).

As long as the decision, if a scientist is granted the requested means, would be solely based on the intrinsic quality of the applicant, in competition with other applicants, the scientist that was successful in that competition still would have its independence in deciding what to investigate. However, most of these programs assess also the "scientific importance" of the proposal, the relation of the requested finances to the size and contents of the proposed research programme, and, moreover and in particular, the societal impact of the project: does the proposal fit with the aimed for societal utilization? (For example the fund may have been set up with the purpose to develop batteries for cars driven with electricity). It is not claimed here that for a society there is something wrong in promoting certain research directions if the result of such research is expected to bring forward the state of welfare of a whole country or the world. But there can be no doubt that such steering of also fundamental research deviates from the main, direct route of scientific progress as determined by the free, independent decision of the scientists. True enough, the scientists following, either forced or yet out of free will, such fund-prescribed research routes, can feel great satisfaction if they see the results of their research applied in practice.

The governmentally financed Max Planck Society of Germany provides the, presumably worldwide, best approximation of complete freedom of research for the Directors of its institutes: the Director is selected on the basis of scientific quality and is supplied with funds of permanent nature to establish extensive research programs only devised by him/her, i.e. without external steering (further see Chap. 13).

The above described science, even if constrained by boundary conditions imposed by the funding organizations, can usually be designated as clean. "Clean science" contrasts with "tarnished science":

In the following three types of tarnished science will be dealt with: foul science occurs (i) if the results of research are bent to serve the interests of the employer or sponsor, (ii) if the provided fund, possibly presented as a gift from a detestable source, is dirty, and (iii) if the professional, scientific behaviour of the scientist involved is immoral. The last example considered below has some connections with my own career and therefore its tale in this booklet is somewhat longer; the narrative also illustrates some moral dilemmas.

Sponsored Clinical Research

Clinical research sponsored by the pharmaceutical industry is an example where the opaqueness of the interaction of the scientists and the industry can hide that the published conclusions of the work are possibly affected by the outcome desired by the industry (*M. Hildebrandt and W.-D. Ludwig, Onkologie, 26 (2003), 529–534*). We should not doubt that at least in some cases that has happened: the *"sponsored studies tend to favour the sponsors' drugs more than studies with any other sources of sponsorship."* (*A. Lundh, J. Lexchin, B. Mintzes, J.B. Schroll and L. Bero, Cochrane Database of Systematic Reviews, 2017, Issue 2, Art. No.: MR000033*; see also, for a similar conclusion regarding veterinary medicine research: *K.J. Wareham, R.M. Hyde, D. Grindlay, M.L. Brennan and R.S. Dean, BMC Veterinary Research, 13 (2017), Art. No.: 234*). The pharmaceutical industry sponsored clinical research provides strong examples favouring the product of the sponsoring industry.

Jeffrey E. Epstein; Maecenas of Harvard and MIT

Jeffrey Epstein was a very successful investment banker. He was extremely wealthy. He was also a convicted sexual offender; his victims were underage girls. At the same time Epstein succeeded to operate as a highly appreciated Maecenas at two high ranking, elite universities: the Harvard University (Cambridge, Massachusetts, USA) and the Massachusetts Institute of Technology (MIT; Cambridge, Massachusetts, USA). Epstein's conviction dates from 2008. This criminal background of Epstein was known at Harvard and MIT. Unbelievable as this may be, this was not a barrier for Epstein to remain a valued person with access to and frequent meetings with high ranking scientists at both Harvard and MIT; Epstein even had an office at Harvard.

The explanation for Epstein's success at Harvard and MIT obviously is his copious financial gifts (millions of dollars). Epstein had contacts with first class scientists, for example Gerard't Hooft, Murray Gell-Mann, David Gross und Frank Wilczek, all Nobel Prize winners, and also Stephen Hawking, Stephen Kosslyn, and many more. He supported at least a few of them financially with considerable amounts of money. Epstein described himself as a "science philanthropist".

Particular problematic is that Epstein was some sort of supporter of eugenics[1] with the crazy idea that the genetic quality of the human race had to be improved by *his* DNA. That steered his donations in specific directions. Epstein had no academic qualifications. Clearly, rich persons can exert with their donations great influence on directions of research, which need not at all be compatible with those which need support the most urgent in the eyes of the scientists themselves.

In 2019 Epstein was again arrested for more of the same as pertaining to his conviction in 2008: sexual abuse of dozens of underage girls, as well as forcing these to prostitution. In the face of a further, much more severe punishment he hanged himself in his cell in prison.

Only thereafter, the scandal being exposed to the public, Harvard and MIT expressed something like apologies and initiated internal investigations. Professor Martin A. Nowak, head of the Program for Evolutionary Dynamics at Harvard has been put on administrative leave. Professor Joichi Ito, the well-known leader of the Media Laboratory of MIT, had to step down. MIT also has tried to avoid identification of Epstein as donor by indicating him as "anonymous". A whole series of prominent, outstanding scientists at both Harvard and MIT had a more than evident hiatus in elementary moral judgement, thereby tarnishing their science.

Abdul Qadeer Khan; Scientist, Spy and Hero; a Father of the Atomic Bomb

In the time that I put my first footsteps in the Laboratory of Metallurgy in Delft there was a tall man, now and then moving through the corridors, with an Indian appearance and name: Abdul Qadeer Khan, of age about 35, who evidently had a close contact with Professor Burgers, but it was unclear to me what he was doing. It turned out that he was preparing a special issue of a scientific journal devoted to the 75th birthday of Burgers in 1972.

Khan came from Pakistan, but was born in India that he left owing to political and religious suppression. After being employed in Pakistan for a few years, he came for study to Europe: he studied metallurgy/materials science at the Technical University of Berlin (at the time West-Berlin) from 1961 to

---

[1] "Eugenics" designates a program to make a certain *inheritable* property more dominant within the human race by reproduction methods.

1965. Then he went to the Delft University of Technology in The Netherlands (called by him: "*the MIT of The Netherlands*") and finished his Master's study in the group of Burgers in 1967. Khan had great respect for Burgers, one might perhaps say that he adored him (Khan about Burgers: "*He was a genius, a thorough gentleman and a good human being*"). Khan stayed an extra year in Delft as research assistant of Burgers. According to his own words, and the reader should remember here that Burgers had one artificial leg: "*During that time I would wait for his arrival downstairs and would help him walk upstairs to his office*". I wasn't already there at the time and cannot confirm it, but it amazes me somewhat to understand that Burgers preferred to take the stairs to his emeritus office at the first floor, which must have been not easy for him, whereas the Laboratory possessed a functioning elevator, of course.

After his stay in Delft, Khan went to the Catholic University of Leuven in Belgium where he performed a Ph.D. project that was concluded in 1972. Burgers together with his wife came all the way to Leuven to attend the defense of the thesis.

I believe it was largely the hard work of Kahn that the special journal issue, a "Festschrift", for the 75th birthday of Burgers was realized: *Journal of Less-Common Metals, 28 (1972), 1–460*. The editors were A.Q. Kahn and M.J. Brabers (Brabers was the promotor of Kahn). The "Festschrift" was handed over to Burgers at a special festivity in the Laboratory of Metallurgy, that I remember only vaguely, as, for some reason, I, as a Master's student at the time, likely did not attend the celebration.

The above text must have made clear that a special relationship existed between Burgers and Kahn. It also suggests that Burgers was not insensitive towards such "adoration" and honours bestowed on him. At least this complies with my own impression of Burgers. Burgers must also have felt grateful and "obliged" towards Kahn.

After having obtained his doctorate Khan returned to The Netherlands and started working for the Physics Dynamics Research Laboratory (= Fysisch Dynamisch Onderzoekslaboratorium (FDO)). FDO, in Amsterdam, was a subcontractor of a firm called Urenco, in Almelo, that developed gaseous ultracentrifuge technology for enriching uranium.[2] Thereby "fuel" for nuclear power plants is produced. However, this product clearly can also be used for the production of atomic bombs.... Obviously, much of the process information was kept secret.

---

[2] A pioneer for this method of uranium enrichment was the Dutchman Professor Jacob Kistemaker. He was a long time director (1949–1982) of the FOM institute AMOLF in Amsterdam, which focused in his time on "atom and molecule physics".

Soon after having joined FDO Khan changed to Urenco. Khan was eventually deeply involved with work on the centrifuges. After India had successfully performed an atomic bomb test in 1974, Khan wanted to help Pakistan in developing the capacity for the production of atomic bombs. It appears Khan gathered a lot of classified information at Urenco about the development of the centrifuges. Suspicion rose in The Netherlands and Khan decided not to return from holidays in Pakistan in 1975. Khan has denied till the present day to have spied and transferred classified information from Urenco to Pakistan.

In 1981, in Kahuta, Pakistan, the site where the atomic bomb program ran, was renamed as the A.Q. Khan Research Laboratories. Kahn acted as its director till 1991.

Khan played a central role in providing countries as Iran, Lybia and North-Korea with classified information about the centrifuge technology: *"The A.Q. Khan nuclear supplier network constitutes the most severe loss of control over nuclear power ever"* (C.O. Clary in his Master thesis, Naval Postgraduate School, 2005, Montery, California, USA). Khan possibly is the person who more than anybody else is responsible for the spread of knowledge on atomic bomb technology. This nuclear proliferation became worldwide news in 2004. The embarrassed government of Pakistan, that Kahn and others claimed to be involved, thought the best policy was to put Khan under house arrest. In 2009 (Khan was 73) this house arrest was relieved. Khan, called "the father of the atomic bomb project", remains a public and popular figure in Pakistan; he is considered as a national hero.

Burgers remained loyal to Khan. In the years after Khan's abrupt "desertion", the Khan affair was big news in Dutch media, leading ultimately to a court verdict of Khan (in absentia) that was later quashed because of a technicality. Khan had no positive press. This did not withhold Burgers from visiting his protégé in Pakistan. Khan bestowed him with another medal: in 1987 Burgers received via a friend of Khan in The Netherlands a large plaquette/medal, a formal honour by the Dr. A.Q. Khan Research Laboratories and The Government of Pakistan according to the text on the medal.[3] This medal is in my possession.

---

[3] This friend is Henk Slebos, who completed his Master project in metallurgy also under guidance of Professor Burgers at the Laboratory of Metallurgy. There the friendship of Slebos and Khan originated. Slebos was twice convicted by a Dutch court, in 1985 and 2005, for exporting goods to Pakistan, presumably for use in the atomic bomb project. Slebos sent the medal/plaquette to Burgers, with an accompanying letter, with the text (translated into English): *"Enclosed you will find a plaquette, especially made for you by our mutual friend"*. It is remarkable that Khan's name is not mentioned explicitly in this letter. This suggests that Slebos thought that (towards a reader of the letter different from Burgers) he might unintentionally draw Burgers indirectly into the Khan affair if he would have mentioned that Khan was a friend of Burgers.

I, unexpectedly, got a puppet role with two appearances in this story. In 1987 I attended an International Symposium on X-ray Powder Diffractometry in Australia. There I was approached by two young scientists from the A.Q. Kahn Research Laboratories in Pakistan. They had been assigned by their highest superior, Dr. A.Q. Khan, to invite me, as a successor of Professor Burgers, to Pakistan and visit Dr. Khan as his guest. I did not like trips to colleagues of a more tourist nature, as I presumed to be the case here, as my job in Delft was imposing. I also did not know exactly how I had to interpret this invitation: I did not know Khan (I had only seen him "passing along", 15 years before; see above), and to develop a relationship with him, a person suspected, at least, to be a spy at the costs of my fatherland, did not seem wise to me. Khan, who after all had lived considerable time in The Netherlands and whose wife was of Dutch origin, had ties with The Netherlands and therefore this was possibly a nevertheless innocent effort to keep in touch. I did not go.

Then, to my surprise, more than 30 years later, in 2018, a second time Khan entered the stage of my life: I got an email sent by Khan (who then was 82), who apparently used the email account of his wife, to my email address in Stuttgart. Kahn wrote that he had enjoyed *"reading my excellent book "Fundamentals of Materials Science"* and referred to the (two) sidenotes I had spent on Professor Burgers in this book. He then recounted his time in The Netherlands ending his recollections with the "Festschrift". He would like to send me a hard copy of the "Festschrift". He finally wanted to know if I would return to The Netherlands (as I had recently retired). I thought the most likely explanation for this email to me simply was the retrospect of an old man on his life (which can also be said from this booklet). Hence I wrote back, informing him that I would stay in Germany and would be pleased to get a hard copy of the "Festschrift". Khan then did send me the "Festschrift" and also his personal hard copy of the Burgers book about recrystallization; he couldn't know that I already possessed the copy used by Burgers himself. These gifts were accompanied by a handwritten letter, where he extensively detailed his memories of the Laboratory of Metallurgy in the mid-sixties. I have to confess I enjoyed reading all this, as it refreshed my own recollections.

Due to grave illness of my wife and problems with my eyes it took about a year before I could bring myself to send an email to Khan thanking him for the Burgers' materials he sent me. That initiated a further email from Khan, with more details about his time in Delft and Leuven, and the surprising remark that he had visited my pre-predecessor in Stuttgart, W. Köster (retired in 1965 (cf. Chap. 13); so Khan likely must have visited Köster after his retirement), to which email I did not respond anymore.

Note added in proof: Kahn became victim of the Covid-19 pandemic. In August 2021 Khan was infected. He died in October of the same year. To illustrate how revered Kahn was in Pakistan, I here simply cite statements made on this occasion: the Prime Minister of Pakistan called him a "national icon" and the President of Pakistan said: "a grateful nation will never forget his services".

Reflecting on the above account, general and personal comments can be given:

In all literature that I read about the "Khan affair", there is no doubt that Kahn literally stole essential scientific and technological knowledge at Urenco, in order to realize the production of an atomic bomb in Pakistan. Such theft is condemnable. Thereby he tarnished his record as a scientist and that of science.

Developing the science allowing the production of an atomic bomb by itself cannot generally be condemned as tarnishing science. Many of us may consider such research, or generally research for the production of weapons, as immoral, but that is not fouling the underlying science.

In my very much younger years I had taken a stance against weapon research. Becoming older I came to accept that such a position was generally flawed. The scientists in Los Alamos working on the Manhattan project (cf. footnote 7 in Chap. 7) were convinced that they had to establish production of the atomic bomb before Germany did. In face of such an existential menace many of us would be in favour of such research; not many of us are that "noble" that they would accept their own extinction and of their fellowmen by weapons of the enemy. So, in this sense, in the absence of durable world peace, i.e. accepting the imperfect morality of mankind, weapon research appears necessary; we should not be naïve. Certainly, if possible, I myself would not want to perform research directly aimed at the development of weapons. But what would I do in a situation as the scientists in the Manhattan project were confronted with? One may compare this dilemma with the one where, on the one side, our desire or even lust for eating meat is met, on the other side, with the recognition that not many of us want to be the one butchering the animals.

Once I visited a heat-treatment shop in West-Germany, where I was shown a very long shaft furnace. The shaft went deep into the bottom of the hall. I was told that its main use was for nitriding the long (6 m or so) cannon barrel of the Leopard tank, a German war machine. This was the first time that I was confronted directly with the application of general results of fundamental research, *where I played a significant role*, in the weapon industry. Of course, our research, on the interaction of nitrogen with iron and iron-based alloys, was of fundamental nature and our intention was not application of the knowledge we acquired in the weapon industry; instead we wanted to understand the "inner workings of nature" (see Chap. 4). If we talked or wrote about the possible technological impact of our work on nitriding, then we indicated the enhancement of the fatigue resistance of machine components, as crankshafts, or the increase of the wear resistance, as of machine components, as bearing parts and the balls in ballpoint pens, or the increase of the corrosion resistance, as of windshield wipers, etc. So we constrained ourselves to mentioning only useful applications which were morally innocent....

As follows from the above, I do not generally condemn scientists working in the weapon industry; we need them to live in peace. In my career it was easy for me to avoid taking part in scientific work of that direct purpose. This leaves unimpeded that I and many other scientists experience the dilemma sketched in the one but last paragraph.

Scientific knowledge must ultimately be available to the entire mankind. It thus cannot be easily prevented that scientific understanding is utilized in also the weapon industry or finds its use also in areas which may possibly be considered as immoral, as, for example, genetic engineering of human offspring. A society that imposes rules to avoid such applications, which are considered as immoral, must be prepared that these rules will be violated. Somewhere and sometime there will always be a "Dr. Strangelove" or a "Dr. Mengele". The scientists involved in the fundamental research at the basis of such knowledge, also necessary for such applications, are not to blame.

The very most of us wish the world devoid of nuclear weapons, atomic bombs, as a war by means of them would kill a large part or even all of us and deteriorate the living conditions on earth in a tremendous way. As this wish is likely unrealizable, the further spread, proliferation of nuclear weapons is attempted to halt by international effort. Then delivering nuclear knowledge and technology, directly leading to atomic bomb production, to dictatorially reigned countries as North-Korea, Iran and Libya, and here Kahn did not deny his active role, is condemnable.

Kahn must have realized that I was more or less aware of his career and ill fame outside Pakistan. It may then surprise that not a single sentence in his emails to me referred to his past as the "father of the atomic bomb in Pakistan". I also did not open such discussion with him: I, presumably correctly, assumed that he, in the last stage of his life, simply needed someone to communicate about those early days at the Delft University of Technology, which must have been for him genuinely unburdened.

The attitude of Professor Burgers is incomprehensible. One would expect that Burgers would have been disappointed about Khan and distance himself from him once Khan had been publicly debunked. The evidence against him was convincing. That did not happen apparently. Burgers went to Pakistan to visit Kahn. I don't know in what way they spoke with each other about the "affair". It may be that Burgers still considered himself as some sort of mentor of Khan, which was his role during the time of Khan in the Netherlands; Burgers and his wife both appeared to have taken, more than in a superficial way, care of Khan and his wife. Burgers never spoke about Khan with me. A letter of response by Burgers to Khan, to thank him for the medal/plaquette he got in 1987, via the route sketched in footnote 3, is unknown. A year later Burgers died.

12

Mistakes, Deceit and Fraud in Science

Abstract "Unintentional errors" versus "deliberate cheating". Even if unjustly, i.e. based on erroneous experiments or thinking, grand claims are made, the border to deceit and fraud in science is not trespassed. Two such cases are presented: the claims of "life on Venus" and "cold fusion". In such cases of supposedly "sensational" findings of potentially worldwide impact, also the "top" journals and the institutes of the scientists involved are too blame by their much too early publication of such findings, which are immature but of expected dramatic impact. ("Self"-)Plagiarism and falsification are first examples of deliberate deceit and fraud in science. The nowadays occurrence of such cases of deceit of, not extremely rare, enormous extent is indicated. Also appropriating for oneself the honour of a discovery in science, actually originating from a colleague, happens, as illustrated for a method now worldwide known unjustly as the "Rietveld method". The chapter ends with a brief account of the possibly most notorious case of sheer fraud in the "hard" sciences: Jan Hendrik Schön invented his data of apparently extraordinary impact and was (thus) able to publish these in an impressive series of papers in the "top" journals.

12.1 Mistakes

Making science is a human undertaking. Making unintentionally errors is human.[1] Those who work as scientists may try to execute their research to the best of their knowledge and capacity, but they will not be flawless. The recorded progress of science testifies of this phenomenon. It is generally believed that science corrects itself: wrong results are weeded out, sooner or later in the course of time, or simply ignored.

---

[1] "To err is human". This statement, presented in this way, is often ascribed to Seneca. But the full citation runs as follows: "To err is human, (and) to persist is diabolical" ("Errare humanum est, (sed) perseverare diabolicum") and the added text, i.e. after the comma, becomes of relevance upon reading the fourth paragraph of this section.

E. J. Mittemeijer, *How Science Runs*,
https://doi.org/10.1007/978-3-030-90095-3_12

Upon establishment of these errors, by the concerned scientists themselves, or others, these deficiencies should be acknowledged by the originators of the inaccuracies, faults and/or blunders. A common way to do that is in an erratum or corrigendum published in the same journal where the defective paper was published. Usually, and correctly, such confessed mistakes, even if not of minor importance, do not necessarily obstruct progress of the career of the scientists concerned in a dramatic way: after all, who is perfect?

The scientists who made the mistakes referred to above are to blame, but their flawed publications were not based on deliberate cheating. Once they admit and correct publicly the shortcomings, further discussion becomes needless.

A difficult situation arises if the mistake concerned involves the cardinal message of a paper. The original authors then may not be inclined to accept and acknowledge that their work is essentially unsound. Admittance of the mistake would imply withdrawal of the entire paper. Indeed, especially if a lot is at stake, one can observe opposition of the original authors. This can happen in particular upon publication (or even before that by press statements!) of apparently sensational results for mankind. Such work draws huge coverage in also nonscientific journals, and broadcasting, television and social media in general, which, by the way, is often sought for by the involved scientists and their institutions. Then it becomes extremely painful to later confess to the world that one was in error, also because thereby one's reputation may be severely bruised.

In the following, two examples of initially supposedly "sensational" findings of potentially worldwide impact, in the above sense, will be briefly discussed. The first example is a very recent one; the second example began with a publication in 1989. The authors of the first, recent case have in fact already acknowledged that their original work is flawed. The authors of the second, old case never withdrew their claims. Notably, in cases as meant here, the evidence presented in work of extraordinary impact, typically is at the limit of what is just detectable by the instrumental tools applied.

It may be said, that extraordinary claims require evidence of exceptional quality. In contrast with this statement, the scientific work in both cases dealt with here appears to be rather imperfect and defective. It seems that the authors and their institutions presented their unripe work publicly and very "loudly" because they believed the sensational nature of their result and as a consequence lost control of what should have been sound experimental work. The untimely run to publicity likely may have had as incentive the hope to be rewarded with highest scientific honour (as a Nobel Prize). These case studies also make clear that not only the scientists concerned are to blame, but also their institutions, which pushed the way-too-early announcements of immature findings, and the journals, which published these papers because of also their sensational quality and the thereby expected high number of citations pushing the impact factor upward.

Life on the Planet Venus?

Venus is our neighbour in the solar system. The planet is of about the same size as the earth (diameter of about 12100 km vs. about 12500 km). It is also a rocky planet and it has an atmosphere. Different from the earth, the atmosphere is constituted of practically only carbon dioxide (CO_2), with a temperature of over 700 K (i.e. more than 400 °C) and a pressure of about 90 atmospheres, both at the surface of the planet. Such conditions do not allow life as we know it on earth. But some decades ago it was speculated that unknown microorganisms might possess properties allowing survival (at high altitude) in the atmosphere of Venus.

It is against this background and especially due to the general public interest for the possibility of extraterrestrial life, that an enormous, worldwide hype was generated by a publication by Jane Greaves and colleagues (the paper lists 19 authors; https://doi.org/10.1038/s41550-020-1174-4, as published by *Nature Astronomy* on 14-9-2020).

The solar spectrum reflected from Venus appeared to show an absorption line at a wavelength where the gas represented by the molecule PH_3, called phosphine, is known to absorb radiation (here from the sun). On earth phosphine is naturally produced by anaerobic life (i.e. life not requiring oxygen) as represented by some microbes, however resulting in a highly variable concentration of phosphine in the earth atmosphere of a factor, say, thousand or more times smaller than the concentration claimed for the atmosphere of Venus in the paper by Greaves et al., which was 20 parts per billion (but see what follows). Phosphine cannot survive under radiation from the sun in the acidic atmosphere of Venus. Thus it was supposed by the authors of the Greaves et al. paper that, in order to establish a concentration of 20 parts per billion, some process must occur which continually stocks up the phosphine concentration. The team around Greaves, after ruling out, in their reasoning, all other possibilities for producing phosphine in the atmosphere of Venus, concluded that, apart from unknown chemistry, the presence of phosphine was indicative of (anaerobic, microbe) life.

The impact of the paper was enormous. However, soon severe criticisms were expressed. To conclude that phosphine is present on the basis of only one absorption line is premature; multiple spectral lines should have been identified. The one line considered in the Greaves et al. paper would not discriminate between sulphur dioxide (a gas known to be present in the atmosphere of Venus) and phosphine.

Scientists reanalyzing the data by Greaves et al. could not find evidence for phosphine. Especially the type of data evaluation may have led to an incorrect interpretation. Greaves et al. used a 12th order polynomial to subtract the background noise around the position where the phosphine line should reside in the spectrum. Such type of polynomial fitting can involve too many adaptable parameters in the fitting of noisy data which can have as consequence the emergence of spurious, i.e. artificial signals. This is a defect well known (also to me and my co-workers, as we more or less daily fitted peak profiles and background in our X-ray diffractograms and other spectra): one can always improve the fitting by adding more variables (using still higher polynomials), but a criterion has to be used to tell the researcher when to stop (which is explained and detailed in textbooks on numerical analysis). Reanalysis of the Greaves et al. data on a sound basis for background noise removal then led to no sign of a phosphine line in the resulting spectrum.

At the moment of writing this text (November 2020), the debate around the work by Greaves et al. has not finished at all. Greaves et al. have produced a follow-up paper (*arXiv.org* > *astro-ph* > arXiv:2011.08176), where they now state that the concentration of phosphine in the atmosphere of Venus is seven times smaller than they originally reported (in September 2020). This cannot be considered as an acknowledgement of having made an unjustified, sensational claim.

One may expect that the discussion on the presence or absence of phosphine in the atmosphere of Venus, and what it would mean (!), may rage for considerable time to come. Ultimately, missions to Venus, likely unmanned, may have to resolve the matter.

Cold Fusion; the Sun in a Pot of Water

Mankind needs energy, a lot of. Most energy sources used till present day are expensive and contaminate the earth, including its atmosphere. Hence a source of energy that is cheap and clean is most welcome.

Nuclear fusion implies the reaction of atomic nuclei leading to the formation of other atomic nuclei, accompanied by the ejection of subatomic particles, as protons and neutrons, and by the release or absorption of energy. Nuclear fusion of *light* elements releases energy. The nuclear fusion of hydrogen leading to helium is the energy source of a star as the sun. The energy radiated, E, is given by Einstein's equation, $E = mc^2$, with m given by the mass difference of the reactants and the products and c representing the velocity of light. The reaction takes place at a temperature of about 14 million Kelvin (about 14 million °C). At this temperature the electrostatic repulsion of the hydrogen nuclei can be overcome (the hydrogen nuclei attain high enough speed at this temperature), the nuclei can approach each other closely, so that the attractive nuclear force can dominate and fusion of the nuclei results.

Against this background of established knowledge, in 1989 Martin Fleischmann (an innovative, highly honoured electrochemist at the University of Southampton, UK) and Stanley Pons (at the time the Chairman of the Chemistry Department at the University of Utah, USA) reported that they had detected fusion at about room temperature, i.e. "cold fusion", in a relatively simple experiment: the electrolysis of heavy water, performed in a pot, leading to the absorption of deuterium in the palladium (Pd) cathode. Deuterium (D) is an isotope of hydrogen (H): hydrogen has a nucleus composed of one proton; deuterium has additionally a neutron in its nucleus. Palladium can absorb large amounts of hydrogen and because isotopes have largely similar chemical properties, the same holds for deuterium. Palladium can absorb a volume of hydrogen gas up to 900 times its own volume. Thereby, the hydrogen or deuterium dissolved in the palladium crystal, there present as ions, H^+ and D^+ (i.e. actually as nuclei only, because H and D possess only one electron), experience a high partial pressure. It was this compression and the high mobility of the hydrogen/deuterium ions in the crystal structure of palladium, which led Fleischmann and Pons to the idea that the deuterium nuclei could come close enough to allow fusion ("*there must therefore be a significant number of close collisions*"). Actually this concept (loading palladium with hydrogen to induce nuclear fusion) was not new, but Fleischmann and Pons were the first to come up with "evidence" that this could happen. The reason that they used heavy water (i.e. D_2O) instead of normal water (i.e. H_2O) is that deuterium upon nuclear fusion generates much more energy.

If the result presented by Fleischmann and Pons would be true, an almost limitless, for practical purposes everlasting source of cheap energy (i.e. sea water) would have become available to mankind.

The paper presenting the considerations and experimental observations by Fleischmann and Pons was published in April 1989: *Journal of Electroanalytical Chemistry, 262 (1989), 301–308*. It was preceded by a press conference on the 23rd of March held by the University of Utah (USA), the employer of Pons. Thereby an agreement with a competing group, led by Steven Jones

(Brigham Young University, Utah, USA), for simultaneous, joint publication, was broken.

The University of Utah had forced this press conference as it wanted to reserve the priority for this "discovery" to itself (see what I said above about "*institutions, which pushed the way-too-early announcements of immature findings*").

It is also remarkable that the paper by Fleischmann and Pons, after been submitted at the 13th of March, was accepted for publication at the 22nd of March. This is highly unusually fast. A refereeing procedure normally takes a couple of months and often much more. It suggests that the expected impact of the paper, that was for sure to capture the imagination worldwide, has led to this high speed acceptance for publication (see what I said above about "*the journals, which published these papers because of also their sensational quality and the thereby expected high number of citations pushing the impact factor upward*").

As consequence of nuclear fusion, protons, neutrons and gamma rays can be generated. Their detection can be used as evidence for nuclear fusion. Jones et al. (see above) focused on the possibly produced neutrons. The particles generated by the nuclear fusion will travel through the palladium and surrounding water and collide with the atomic/molecular entities in their environment. Thereby heat is generated. It is this heat that Fleischmann and Pons largely focused on in their experimental work. They claimed that much more heat was produced than could be explained by chemical reactions alone. This excess heat (i.e. energy) release, they stated, had to be ascribed to unknown nuclear reactions.

The impact of this news was gigantic. The credibility of the authors (in particular Fleischmann had a high status) at first supported this effect. Especially the supposed availability of a gargantuan source of cheap energy (see above) stood central, as in newspaper announcements.

Also in my group in Delft this news was amply discussed during coffea break. The journal with the Fleischmann and Pons paper was well known and easily available to us in our laboratory library (electronic publications and electronic libraries did not exist already). So we could read the paper more or less at the moment it appeared in print. Our interest also arose because in the past, under Burgers' guidance, research had been executed on the absorption of hydrogen in palladium. Moreover, because of our own running nitriding research, we were well aware of the high partial pressure of a dissolved gas in a metal: nitrogen (N) dissolved in iron (Fe) can be associated with a partial pressure of up to 10^5 atm[2] and it also has a high mobility. However, this does not lead to nuclear fusion, but involves the strong tendency for the dissolved nitrogen to recombine as nitrogen gas molecules (N_2) and precipitate as pores within the metal, which is also a peculiar phenomenon, studied by us intensively many years later in Stuttgart.

Within weeks after the announcement by Fleischmann and Pons, criticism started to develop. Many experiments by many other scientists were performed to reproduce the findings of Fleischman and Pons. They generally failed. Several times the status of the field was reviewed by scientific committees, but soon the overall opinion emerged, and remained as such, that the experimental claims and the interpretation by Fleischman and Pons could not be upheld. Moreover, various experimental flaws and shortcomings in the experiments by Fleischmann and Pons were exposed. Yet, there is no reason to think that they manipulated or invented their data.

In this case, the scientists involved, Fleischmann and Pons, never withdrew their claims and more or less remained stuck with "cold fusion research" that they continued for some years at a laboratory in France.

Nothing in science can be said with truly absolute certainty. Indeed there is a modest group of scientists worldwide that continue research on cold fusion, more or less of the type considered here; they also get funds for that. We must always be prepared to adapt our model of nature as soon as validated, experimental results are presented which deviate from what we at the time think to be nature. Our model of nature is only *for the time being* an appropriate description of nature (Chap. 4).

As a final note it cannot be avoided to point out that an unassuming approach in presenting one own's results is advisable. As a striking example how a sensational new piece of science can also be presented, the discovery of high T_c superconducting materials by Bednorz and Müller in 1986 can be mentioned (T_c denotes the temperature below which the material concerned is a superconductor, i.e. the electrical resistance has dropped to zero). They published their results in a good journal, but not in a journal of highest impact (as, for example, *Nature* or *Science*), notably: *Zeitschrift für Physik B, Condensed Matter, 64 (1986), 189–193*. Also no press conference was held to underline the sensational nature of this work.... The work had a large

---

[2] Interestingly, whereas it has been claimed by Fleischman and Pons that the dissolved deuterium would experience a partial pressure of the order 10^{26} atm, it was stated that, owing to a calculation error made by them, the partial pressure actually was of the order 10^4 atm (see *J.R. Huizenga, Cold Fusion: The Scientific Fiasco of the Century, 2nd edition, Oxford University Press, 1993*), as holds for the dissolved nitrogen considered here.

impact on science and its potential application: superconductivity at temperatures higher than 77 K (= -196.2 °C; the boiling temperature of liquid nitrogen) allows in principle engineering applications of superconductivity. Soon the results by Bednorz and Müller were reproduced and T_c temperatures distinctly above 77 K were obtained. I well remember that, very soon after the discovery by Bednorz and Müller, as at many places on the world, also at the Philips Research Laboratories (NatLab) in Eindhoven, under the leadership of Andries Miedema (see Chap. 13), a "task force" was formed with the aim to find compounds of ever higher T_c, by means of, what we, as outsiders, called unrespectfully: "cooking and baking". Bednorz and Müller were awarded with the Nobel Prize in Physics of 1987. Remarkably, the origin of the superconductivity in high T_c materials is still unclear.

12.2 Deceit and Fraud

Whereas the border, separating what is (just) acceptable from what is indecent and possibly even criminal, has not been crossed for the cases discussed in Sect. 12.1, traversing this frontier is dealt with in this section.

A very long time ago the naive impression of the general public might have prevailed that science is an activity performed by people who are perhaps not flawless but who perform their assignments very carefully and accurately: scientists could be trusted.

The first severe damage done to this image may have been the proven occurrence, especially in the last decade, of the devastating lack of reproducibility of very many results published in particularly the social and psychology ("soft") sciences. Reproducibility rates of (far) less than 50% occur for example in psychology. In case of social and behaviour science papers published in even top journals as, for example, *Nature* and *Science*, the results of more than 1/3 of these papers were found to be irreplicable *(C.F. Camerer et al., Nature Human Behaviour, 2 (2018), 637–644)....* In medicine it is not better *(J.P.A. Ioannidis, PLoS Medicine, 2 (2005), 0696–0701)*. In the corresponding literature one speaks of a "replication or reproducibility crisis".[3] This may not have amazed many of us in the natural ("hard") sciences, but to recognize that such irreproducibility could also, but less frequently, be

---

[3] A fine difference between the notions "reproducibility" and "replicability" may be noted. "Reproducibility" can be defined as the possibility to duplicate the results of a prior study using the same materials and procedures as in the original study. "Replicability" is less stringent as "reproducibility", as it requires that the results of a prior study can be confirmed applying the same procedures as in the original study, but for different data, and thereby producing similar results. This distinction does not play a significant role in the present discussion.

connected with apparently serious publications in the natural sciences was more difficult to digest.

These irreproducible results for a large part may have not been the result of deliberate cheating, but may have been the consequence of careless experimental work and sloppy interpretation (cf. Sect. 12.1). However, the number of papers, presenting, in full awareness of the authors, wrong or manipulated data and results, appears to increase rapidly and seems to originate especially from countries not at the forefront of science (see below). That then is cheating.

In retrospect, deceit and fraud have been of all times in also science. The difference with the past is that there appears to be so much more of it nowadays. This has only partly a statistical background, as due to the relatively very high number of currently living and active scientists (cf. Chap. 6). For sure this is also connected with an on average changed, less noble attitude of many of the present-day scientists with respect to science as a profession and calling (cf. Chap. 6).

Self-plagiarism and Falsification

The first time I was directly confronted with a case of shameless duplication of a scientific paper occurred during my stint as Editor-in-Chief of the *International Journal of Materials Research (IJMR)*.

I was approached in 2017 by the Editor of the *Journal of Nanoparticle Research (JNP)* informing me that a paper published in *IJMR* by Masoud Karimipour et al. with the title "Rapid synthesis of Ag nanoparticles and Ag@SiO$_2$ core–shells", *(IJMR, 106 (2015), 532–534)* was very similar to a paper published in "his" journal *(JNP, (2015) 17: 2)*. The only significant differences were a different title, "Microwave synthesis of Ag@SiO$_2$ core–shell using oleylamine", and the addition of a further name to the list of authors. It appeared that Karimipour et al. had submitted practically the same manuscript, apart from the mystifying change of title and the change in the list of authors, at the same time to both *IJMR* and *JNP*. I felt obliged to immediately publish a "Note from the Editor-in-Chief" in IJMR making this case of unacceptable duplication clear to the readership of *IJMR*.

Hoping that this unpleasant experience would be a singular event in my career as Editor, I was soon to learn the opposite: much worse was to come.

In 2019 I was approached by a "whistleblower", informing me that three papers published in *IJMR* in 2011 (from before my time as Editor-in-Chief) contained, partly or largely, duplicated and falsified material from other papers published in other journals by the same leading author, Ali Nazari *(IJMR, 102 (2011), 457–463, 560–571 and 1312–1217)*. We performed a check of the literature by ourselves and could only confirm the claim by the whistleblower. In my ensuing published "Note" to the readership of *IJMR* I wrote:

"These papers published in IJMR have significant overlap in terms of identical content and wording with papers published by Ali Nazari et al. in other journals; strikingly the same micrographs and numerical data were used in different papers, albeit discussing different materials (additives). Examples of such papers published in other journals include (and thus are not limited to): Journal of Composite Materials 45(8) (2010) 923–930; Materials Research 14(2) (2011) 178–188; Energy and Buildings 43 (2011) 864–872."

The deceit by Nazari et al. has much larger dimensions than indicated by the above. The case is described on the website of "Retraction Watch" (https://retractionwatch.com/; parent organization: Center for Scientific Integrity).

Regarding the papers from this author, we are confronted with not only duplication of papers, but also with falsification: scale and contrast of images were varied, different parts of the same micrograph (possibly in rotated fashion) were published in different papers; the same micrograph in different papers was stated to represent different materials in different conditions; similarly for X-ray diffractograms. Rather different lists of co-authors were presented for apparently the same work.

At the time of publication of the retraction by *IJMR* already 27 papers by Nazari et al. were retracted by other journals. Perhaps surprisingly, a number of journals were slow in retracting, if at all, the duplicated and falsified work.

Ali Nazari was employed by the Islamic Azad University in Iran and later by the Swinburne University of Technology (Australia). He now has lost this last position. Nazari has authored more than 200 papers. The question is justified if the "work" presented in these papers has ever been done.[4]

It even appears that four groups of Iranian authors (including "Nazari's group") must have communicated and cooperated in the type of misconduct, as image sharing, as described above.

---

[4] The enormous extent of the deceit by Nazari et al. is less exceptional than one might think. In 2010 the Editors of *Acta Crystallographica* disclosed that at least 41 and 29 papers published in the journal by two groups in China, respectively, had to be retracted (*Acta Crystallographica, E66 (2010), e1–e2*). The Editors wrote: *"… a bona fide set of intensity data, usually on a compound whose structure had been correctly determined and reported in the literature, was used to produce a number of papers, with the authors changing one or more atoms in the structure to produce what appeared to be a genuine structure determination of a new compound.".* Evidently, this also is a further illustration that the peer evaluation system applied to submitted manuscripts can fail dramatically (also see Chap. 6). In the case considered here the Editors further wrote: *"… chemically implausible or impossible structures arose from these manipulations, and it is a concern and disappointment that these chemical features passed into the literature undetected.".*

A number of remarks now are in order:

- The exposure of this large scale of grave scientific misconduct is the work by a single person, a "whistleblower". The journals which published the "work" by Nazari et al. and related groups have failed to reveal the duplication and falsification in their refereeing stage. Very many referees must have been involved; they did not notice the fraud done before their eyes.
- The field of science pertaining to the "work" by Nazari et al. deals with construction engineering materials. Thus it is of strongly *applied* nature. It may be speculated that fraud on the scale as discussed here is less likely possible in a field devoted to *fundamental* science, where a paper may have a wider significance (but see the last case study in this chapter). The scientific merits and lifetime of a fundamental paper may be larger. In any case, the field of construction engineering materials, experiencing an exponential, very steep increase of the number of published papers, seems to be damaged by a huge burden of low quality and even fraudulent papers.
- I have not refrained from indicating the origin/nationality of the duplicating and falsifying authors in this section: Iran. Iranian authors appear prominent in surveys of duplicating and/or fraudulent papers (see "Retraction Watch"). This does not at all mean that this holds for all science produced by Iranian authors. But an apparent relation between scientific misconduct and country of origin/nationality should not be suppressed (also see footnotes 10, 11 and 12 in Chap. 6). Moreover, of all manuscripts, by Iranian authors and from Iran, submitted to *IJMR*, the rejection rate is very high (more than 80% for the last 5 years), which indicates the on average low quality of the submitted work. It is true that the latter observation can also be made for some other countries with a less developed scientific tradition.

The Rietveld Method; an Improper and Dishonest Namesake

In physics, chemistry and materials science the determination of the crystal structure of a compound, i.e. the determination of the atomic coordinates, for all types of atoms in the compound, is usually performed by (X-ray) diffraction methods. The classical method is to take a large enough single crystal and to measure the "positions" (i.e. as determined by the orientation of the single crystal and the direction of the diffracted beam) and the intensities of as many as possible reflections. With "intensities" here is meant the total intensity contained in a reflection, i.e. the area under the peak profile representing the reflection, which is usually called "integrated intensity". Usually the receiving slit width is chosen that large that the entire reflection is encompassed and the only intensity value recorded for each reflection then is the integrated intensity (after background elimination). Starting from a model crystal structure, the atomic parameters, as their coordinates, then can be refined in a fitting procedure with the criterion that the integrated intensities must be well fitted.

Often large enough single crystals are not available. In such a case one can take a so-called powder specimen of the compound concerned, i.e. a very large collection of (very) small crystals of the compound. A diffractogram ("diffraction pattern") of this powder specimen can be recorded. Each reflection in this diffractogram has contributions of those crystals in the powder specimen which happen to be in an orientation allowing diffraction. The reflections in such a diffractogram can severely overlap, which obstructs a direct determination of the integrated intensities. For the "powder diffraction method" this disadvantage for a long time has been the major hindrance to its application for crystal-structure determination.

The solution for this problem came from scientists at the Reactor Centre Netherlands in Petten, where neutron diffraction was applied for resolving the atomic structure of compounds, as well as their magnetic structure.[5] Both X-ray and neutron diffractograms of powder specimens suffer from the overlap of reflections. In the first half of the sixties of the past century, Bert O. Loopstra, a tall and lean man of modest personality, proposed, also on basis of discussions with his colleague Bob van Laar, *to fit the entire powder diffractogram, comprising all measured peaks as profiles, as an entity,* by refining the parameters describing the crystal structure *and* the parameters describing the peak shape, where the peak shape was in principle different for each reflection. This implies that not integrated intensities are fitted but every measured, individual "line intensity" (experimentally a peak thus is composed of a number of adjacent line intensities) is fitted in a least-squares procedure.

The idea was applied in practice after a scientist with computer and numerical computation experience was added to Loopstra's group. The first paper where the "profile refinement method" was applied and presented appeared in 1969 *(Acta Crystallographica, B25 (1969), 787–791)*; the paper was submitted on the 17th of April 1968. A later paper *(Journal of Applied Crystallography,*

---

[5] Neutrons possess a magnetic moment, which implies interaction with the magnetic moment due to the electrons surrounding the atomic nuclei. A further difference with X-rays is the much large penetrative power of neutrons, so that information from larger depths beneath the surface of a specimen can be obtained.

2 (1969), 65–71) was submitted on the 29th of November 1968. Perhaps unexpectedly, it is only this later paper that is usually referred to as the origin of the "profile refinement method". It has led to the name "Rietveld method" or "Rietveld refinement", as Hugo M. Rietveld, the scientist with computational experience referred to above, had submitted that paper with himself as the only author; Loopstra and van Laar were mentioned in an acknowledgement for *"suggestions and helpful criticism"*. This did grave injustice to the originator of the method, Bert Loopstra (Loopstra and Rietveld had been the authors of the very first paper applying and presenting the "profile refinement method"; see above).

Up till about 1973 the "Rietveld paper" was not very much referred to. Only after 1975 the citation rate started to increase significantly: after more than one computer program, from diverse groups, for application of the method, had become widely and freely available and a first book on the method had appeared in 1993 *(R.A. Young (Editor), The Rietveld Method, Oxford University Press, 1993)*, the number of citations increased pronouncedly, now (in 2020) having reached a total number of about 13000. Likely most of the authors who currently refer to this paper may not have read it.

Nowadays the "profile refinement method" is not only restricted to crystal-structure determination/refinement. Whereas in the original method a profile-shape function of Gaussian, pseudo-Voigt or Pearson VII type, etc. is employed, with profile parameters of numerical value in principle different for each reflection and to be determined in the fitting of the entire diffractogram/diffraction pattern, the profile shapes can alternatively (better, i.e. physically more founded) be described by microstructural parameters, as the crystallite size and the mistake density, which then can be refined as well. Also residual (macro)stress and preferred orientation parameters can be determined and the method can also be used for quantitative phase analysis. Thereby the "profile refinement method" has become an important tool for microstructural characterization in especially the field of materials science as well. This paragraph may serve to make likely how enormous the impact of the "profile refinement method" has become.

Rietveld's colleagues in Petten did not appreciate his conduct in handling scientific credits. In this context one may interpret Rietveld's move to the position as librarian at the Reactor Centre Netherlands in 1974.

In the same year I attended the Second European Crystallographic Meeting (ECM2) in Keszthely, Hungary (26–29 April, 1974). I was 24 and in the middle of my Ph.D. project. Rietveld attended the meeting. He was often surrounded by a larger group of conference participants and apparently was able to draw the attention. It was said he had very recently accepted the position as librarian because "his method" had not found the wide acceptance as he had expected and now "disappointedly" left science. This fairy tale is still believed by some in the field of crystallography, as evidenced by an obituary (Rietveld died in 2016). However, at the same meeting I also picked up that there had be a "discussion" about "priority" and authorship at Petten and that Rietveld didn't have "clean hands". It was and remained clear since then, to me, and at least a few other Dutch crystallographers, that Loopstra deserved, more than Rietveld, credit for the "profile refinement method".

I came to know Bert O. Loopstra (who, after his time in Petten, had become Professor at the University of Amsterdam) personally, because his son Onno B. Loopstra (note the amusing reversal of the first characters of the Christian names of father and son) did his Ph.D. under my guidance (1992), at a stage where Bert Loopstra had already retired. Bert Loopstra never mentioned to me his involvement in the "profile refinement method" and the injustice done to him. He died in 1998.

The "Rietveld method" is one of the most distinct advances in crystallography of the last 50 years or so. In the course of years the scientific world bestowed many high honours on Rietveld. Rietveld presented himself more and more as the sole inventor of the method and did not shy away for manipulations in that direction. A number of these ugly, but revealing details have been recounted by my colleague Professor Henk Schenk, a former President of the International Union of Crystallography, IUCr (the IUCr is publisher of the journals *Acta Crystallographica* and *Journal of Applied Crystallography*, wherein the original papers on profile refinement appeared) and Bob van Laar, who contributed significantly in the development of the "profile refinement method" *(Acta Crystallographica, A74 (2018), 88–92)*.

Evidently, the "profile refinement method" should more justly be called "Loopstra method". However, after the name "Rietveld method" has been such generally introduced and adopted as is the case now, a renaming appears an impossible undertaking. Thus this story has a sad ending: deceit in science, by appropriating an original idea that is not one's own, can be very worthwhile for the perpetrator.

Sheer Fraud; the Infamous Jan Hendrik Schön

If scientific fraud concerns a seemingly important result of moreover great impact, then it is for sure that the fraud will be uncovered: many scientists all over the world will embark on research trying to reproduce the claimed phenomenon and to extend the work, aiming at further advance and, possibly, enlargement of the range of applications.[6] Since these efforts will be in vain, i.e. basically no reproduction is found to be possible, the swindle will soon be exposed. Hence, this type of fraud is extremely rare.

However, if scientific fraud concerns falsified work that, in view of its message, certainly is of trifling importance, it may remain unnoticed, because the papers concerned are not seriously considered. Indeed, most of the published scientific papers don't play a pronounced role for the progress of science and may not even be cited once (cf. Chap. 6). Thus a fraudulent nature may remain unnoticed for a long time, if not for ever. The large number of fraudulent papers published by Ali Nazari (see above under *Self-Plagiarism and Falsification*) serves as an example: the discovery of the falsification underlying these papers was only due to the persistence of a whistleblower. It can thus be presumed that many of such fraudulent papers remain unnoticed, as the work is simply ignored as being in any case irrelevant.

As *the* example of the extremely rare type of fraud where great discoveries are claimed, the vast deceit committed by Jan Hendrik Schön is discussed here. It concerns the probably most notorious, shocking case of fraud in the fields of physics and materials science ever. Extensive literature about it has been published after its showdown; even a book has been produced *(E.S. Reich, Plastic Fantastic: How the Biggest Fraud in Physics Shook the Scientific World, Palgrave MacMillan, New York, 2009).* The choice of this example has been motivated by my profound bewilderment about a culprit who evidently expected to get away with it and also by a slight personal connection.

After completing his Ph.D. in 1997 at the University of Konstanz in Baden-Württemberg (i.e. in the extreme south western part of Germany), Schön accepted a position as a post-doctoral staff scientist at the famous Bell Laboratories in the USA (cf. Chap. 5). His task was to find an alternative for the silicon-based semiconductor materials on the basis of organic compounds. It was later shown that already in his first paper from his time at Bell Laboratories, submitted in 1998 to the *Journal of Applied Physics*, manipulation of data had been performed in the sense that additional "experimental" results, not based on any performed experiment, were added to a figure. This was just an upbeat for more serious fraud coming next.

Schön thus "discovered" the quantum Hall effect and superconductivity in organic compounds, developed a field-effect transistor based on one layer

---

[6] Obviously, the same verification mechanism in science also operates if no fraud but mistakes are the origin of unjustified, grand claims in scientific work, as discussed for the examples in the first part of this chapter.

of organic molecules, the first organic laser and, as a crowning result, the first field-effect transistor based on a single molecule. Schön's results were prodigious. His productivity was extreme; he published 45 papers in 2001, i.e. one paper every nine days, with an apparent quality such high that four papers appeared in *Nature* and four in *Science*, which are top journals. People started to talk about the Nobel Prize waiting for him.

Top journals select papers also on basis of expected extraordinary impact. This certainly explains partly the success, i.e. acceptance for publication, of the "work" by Schön at specifically these journals. On one occasion *Science* accepted a paper by Schön within three weeks, whereas normally a number of months, and often much more, are the rule (see under "*Cold Fusion; the Sun in a Pot of Water*" in Sect. 12.1, and, in particular, the discussion under "*Editor of a materials science journal*" in Chap. 6).

Already in 2001 people got suspicious about Schön's scientific claims. When requested, Schön did neither provide his original samples nor his raw experimental data for other groups to perform (reproducibility) measurements. In May 2002 it was demonstrated unambiguously that exactly the same set of data was used in two different papers (one *Nature* paper and one *Science* paper), although the context of the data was different, i.e. the same data would refer to apparently different experiments. Also the noise attached to his data in different papers was found to be identical for apparently different experiments. There was no doubt anymore: Schön had fabricated the experiments. A committee was founded by Bell Laboratories. The conclusion of its investigation was a long list of cases of fictional and duplicated "experimental" data in a series of papers which in the end had to be withdrawn. Schön was fired immediately. He, who undoubtedly must have known what he was doing, never admitted more than that he had made a number of mistakes in his work....

As these events ran, I was Director at the Max Planck Institute for Metals Research in Stuttgart. Since February 2002 this institute was housed in a new building tightly connected with its sister institute the Max Planck Institute for Solid State Research: floors at the same level; walking through the corridors one could enter or leave the one or the other institute; also common facilities and services were joined: one library and one canteen for both institutes and joined technical support services. This serves to explain facile contacts between also the Boards of Directors of both institutes. One was well aware of intended larger changes in each other institute, as associated with the search for candidates for vacant Director positions. Klaus von Klitzing, who had obtained the Nobel Prize of 1985 for the discovery of the quantum Hall effect, had invited Schön for a visit to his Institute for Solid State Research in already 2000. In the end this institute had the intention to propose that the Max Planck Society should offer Schön a Director position at the institute. After Schön was exposed publicly in 2002 as a deceiver, before such a Director appointment had actually been realized, a "sigh of relief", that such embarrassment had not touched the institute, could, so to speak, be "heard and felt" till within my office at the sister institute.

There is a peculiar aftereffect to the downfall of Schön: The University of Konstanz started an investigation of Schön's Ph.D. thesis of 1997. No fraud could be found. Yet the university withdrew the granted "doctor" title in 2004, with the reasoning that Schön had shown to be unworthy to carry the "doctor" title. A law of the Land Baden-Württemberg (Germany is a federation) made this possible. Schön appealed and won his case in 2010. Now the university appealed. Finally Schön definitively lost in court in 2011: the court concluded that "grave scientific misconduct" justified the imposed sanction of title loss, i.e. also in the absence of a criminal act.

In view of the grandness of his scientific claims, it amazes profoundly to observe that Schön expected to remain unexposed (see above). However, as indicated by Reich (the author of the book referred to above), deceivers as Schön may yet be considered as "first" discoverers of a certain effect if subsequent researchers can confirm the claim, but then on the basis of true data. Here it must be noted that the claims by Schön fell within what the "world" was looking for: electronics on the basis of conducting organic materials and the development of devices based on the action of a single layer of molecules or even a single molecule, which suggested the future development of computers of high(est) power at very low cost. Also more sophisticated manipulation in his papers would have been possible: adding some outliers and more skillful fabricated noise, avoiding the direct duplication, would have made discovery more difficult and likely led to a very much later exposure of the fraud. These are disturbing thoughts.

The institutions, where scientists work, must take care that no scientific misconduct by their employees takes place. After various rather recent cases of scientific misconduct of variable character, nowadays awareness for introducing rules for storing the raw data of experiments and for openness for criticism of colleagues independent of hierarchical structures ("internal screening"), and many more, are a prerequisite to avoid derailments of serious nature. As an example, the "Rules of Good Scientific Practice", as issued by the Max Planck Society, may serve.

Finally, the responsibility of the top journals cannot be marginalized. They are lured to publication of immature, even fraudulent work, as long as the instantaneously exciting, "capturing-the-imagination" character of the work guarantees a huge response, as shown in this chapter by examples. This will yet lead to increase of the impact factor of these journals, already only by citations in the following, refuting papers and other publications, but that does not reflect a corresponding increase in quality of the journals concerned.

13

The Stuttgart Years;
Science at the Max Planck Institute
and the University

Abstract Start and first experiences in Stuttgart. My personal reminiscences of my "predecessor" Professor Predel. A brief history of the Max Planck Institute for Metals Research; processing of its Second World Wartime past in 2002. Move to the new building in 2002. Differences between manners and mores in "Delft" and "Stuttgart"; experiences with neighbours in private life. The unexpected restart of nitriding research, initiated by the discovery of amorphous nitride precipitates. The spectacular role of surface and interface energies in transformations at surface and interfaces; the model to predict, for real materials, quantitative values for these energies. The elimination of long held and widespread misconcepts: why the natural oxide layer on a metal (as aluminium) is amorphous and why whiskers are formed on a thin (tin) layer. Dean of the Study Course Materials Science; the "Kollegialprüfung". The ambiguity of the relationship Max Planck Institute—University. Drafting the book "Fundamentals of Materials Science". Colossal, internal stress gradients in thin films. Discovery of oscillating stress in a thin film: a quantum mechanical surprise of nature. The chapter ends with an example illustrating the perhaps greatest satisfaction for a materials scientist: the experience that his/her research of processes occurring in and with materials leads to fundamental, physical understanding of nature that allows utilization in technologically important applications.

The plane had begun its descent. Eric looked out the window. He saw the sloping hills surrounding Stuttgart; the visual impression was of green patches surrounded by forests. He liked what he saw. He always had considered the flat Dutch landscape as rather dull, non-exciting. He especially well remembered the Sunday walks he had to make as a small boy with his parents in the "Haarlemmermeer polder", south east from Haarlem: unending straight roads, without any up or down, with at both sides endless open fields. These monotonous strolls were an extremely boring experience. Perhaps this was the explanation for his later love, as a mature man, for mountain hiking.

© The Author(s), under exclusive license to Springer Nature
Switzerland AG 2022
E. J. Mittemeijer, *How Science Runs*,
https://doi.org/10.1007/978-3-030-90095-3_13

Closing in on Stuttgart the density of villages and small towns increased. A lot of motorways, overloaded with traffic, cut through the landscape. This was a signal that should have warned Eric that in Germany Stuttgart was a city with most traffic jams on its supply roads; to his annoyance he would have to live with this phenomenon for about the next 20 years.

Shortly before touchdown one could observe that Stuttgart lies in a bowl: the main railway station and the city centre are located at about the lowest level; the main streets starting there all ascend to the rim of the surrounding hills, about 100–200 m higher, where suburbs can be found. Eric immediately decided that the place to live should not be near the bottom of the bowl, but preferably in a higher-up suburb: he could imagine that air pollution would make living in the bowl unhealthy.

Stuttgart is not beautiful; its surroundings are. The centre of the city had been devastatingly bombed in the Second World War. Practically no building of some architectural significance of past centuries had survived. The rebuilding of the city in the fifties has provided the city with a mass of ugly buildings and an appearance of the city centre that cannot compete with, for example, that of Munich. At the time recounted here, Eric had lived for many years in Delft, a small city with a beautiful medieval center and an atmosphere impregnated by the university. For a German reader: Tübingen, close to Stuttgart and also a small university city, comes close to Delft. The loss of such a "biotope" would be regretted severely by Eric.

He had never been before in Stuttgart. This early morning he now had come to present himself to the Max Planck Institute for Metals Research (MPI-MF) and the Faculty of Chemistry of the University of Stuttgart. In a way, this morning, with its feelings of uncertainty not knowing what lay ahead of him, awaked memories of long ago, when he had come to Delft for the first time, as a young man of 17 years, to enroll as a student at the Delft University of Technology.

At the airport Eric was picked up by Mr. Schlenker. Mr. Schlenker was an employee of the MPI-MF, who was of distinct importance to the institute, since he, as a genuine handyman, a designation meant absolutely respectfully, generally solved many operational, technical problems and, as Eric was to find out, would also be of great support to him personally and his department. Mr. Schlenker, who was the contrary of talkative and in a sympathetic way always kept the distance between him and a Professor/Director, was one of the finest characters Eric would experience during his Stuttgart years.

Schlenker drove Eric from the airport to the institute, down town in the Seestrasse. A strenuous day had begun to unfurl.

Start in Stuttgart

The position in Stuttgart was a special one: a Director position at the Max-Planck-Institut für Metallforschung (= Max Planck Institute for Metals Research; MPI-MF) in conjunction with a full Professor position (Chair Holder) at the University of Stuttgart in the Faculty of Chemistry. This construction had great appeal to me, as it allowed access to students and participation in the study course Materials Science (at the time indicated in Stuttgart by the name in German "Werkstoffwissenschaft"), which was part of the Faculty of Chemistry. I strongly believe in the fruitful interaction of students and active scientists and that would be less well realizable as only a Director at the MPI-MF, apart from the fact that I love to teach. This explains why the "nomination committee" was composed of both Max Planck Directors and Professors of the University.

The center of gravity of my first day in Stuttgart was a public lecture and a following discussion in a closed meeting with the nomination committee. I had chosen for the lecture the simple title "Stickstoff in Eisen" (= "Nitrogen in Iron"). This topic allowed illustrating the versatile, fascinating and striking behaviour of nitrogen (N) in iron (Fe) and spanned topics from experimental to theoretical character, thermodynamical and kinetical analysis. Thus I expected to make clear that I was the scientist appropriate for the position available. I had decided to lecture in German. This had the advantage that a conceivable criticism, that I, as Dutchman, would possibly not be able to communicate in German with students in Stuttgart, would have evaporated before expressed.

On this day and the following one, the institute was shown to me. Part of the institute was located rather close to the centre of Stuttgart in the Seestrasse, where I would be located; another part in Büsnau, a suburb of Stuttgart. I met my future colleagues. Most of them had a name that meant something to me. Thus I knew work done by Alfred Seeger and Helmut Kronmüller, but I had never met them in person. Seeger had retired rather shortly before. I had an enjoyable discussion with him and later learned that he had strongly supported my nomination.

Professor Seeger had not only a very high scientific reputation, but also was known or perhaps even feared because of his sharp tongue. Many years earlier, my colleague in Delft, van den Beukel (see Chap. 10), told me that he once, in the beginning of the sixties, had presented work from his Ph.D. time at an international conference. Van den Beukel's Ph.D. thesis (1962) was devoted to the recovery stages of deformed metals as revealed by the release of energy (heat) upon annealing. This was a topic about which Seeger and coworkers had extensively published in preceding years up to the time of the conference. At the meeting a "collision" in public between van den Beukel and Seeger occurred, as their interpretations did not match, which encounter unfortunately got strongly personal. This caused that, many years later, but long before my leave to Stuttgart, in a rambling discussion van den Beukel recalled this episode towards me expressing his negative impression.

Further, along other routes, I had been informed about some "friction" in the past within the MPI-MF associated with the hierarchical organization of the Board of Directors at the institute. For a long time the MPI-MF was split into two subinstitutes: the "Institut für Werkstoffwissenschaft" (= "Institute for Materials Science and Engineering") and the "Institut für Metallphysik" (= "Institute for Metal Physics"). These subinstitutes had two *permanent* "Managing Directors": Prof. Hellmuth Fischmeister and Prof. Alfred Seeger respectively. This implied that the other Directors at MPI-MF had a "boss", either Fischmeister or Seeger. Shortly before I seriously considered my move to Stuttgart, the organization was drastically changed: from 1997 on all Directors got the same weight; one of them would be acting Managing Director, for a period of two years, to be succeeded by another Director thereafter, etc. The authority of the Managing Director was basically reduced to executing the decisions taken by the full Board of Directors. I let this explain in detail to me by the momentary acting Managing Director, Manfred Rühle, during the time of my negotiations with the Max Planck Society (MPS). Against the above described background and the independence I had and cherished as a Chair Holder in Delft, in no way I would have accepted a position with a colleague as my permanent superior.

Part of the negotiation procedure was a meeting with the President of the Max Planck Society: Hubert Markl, who served as President from 1996 till 2002. The head office ("Generalverwaltung") of the MPS is in Munich. At the time the President was especially safeguarded (it was the aftermath of the period of RAF terror) and one could not use the elevator up to the floor of his office without more ado. I had not realized that the President of the MPS was one of those high ranking officials in the Bundesrepublik who was considered a potential victim of terror; the precautions were a special experience.

The discussion with Markl was a very pleasant one. He asked me to tell him what drove me in science. I spoke about my fascination for the mobilities of the constituents of matter, atoms, which lead to distinct microstructural changes in solid materials, as metals, in many cases already at room temperature: observed from the outside a piece of material, for example a steel or an aluminium alloy, can be at rest, no sign of any changes in its internal structure, which yet take place, with pronounced consequences for the properties of the material. Why does this happen ("thermodynamics") and how do the atoms move ("kinetics") can be seen as the questions powering my research. Also the role of internal stresses was highlighted: I remember I explained what kind of internal stresses were acting within the table at which we were sitting… and how I was generally intrigued by the interaction of internal stresses and mobilities of the atoms. Markl was genuinely interested.

Seven years later we more or less "bumped" on each other on a street in Stuttgart. We had not met since our first meeting in Munich described above. Yet he immediately recognized me, knew my name (although there are more than 230 Directors of the MPS) and asked me about my activities and especially how I felt at the Institute. Markl was a sympathetic man and an impressive yet modest personality (also, see later in this chapter).

As my position in Stuttgart involved two jobs at the same time, I had to negotiate with the University as well. In that case money and personal positions were not the major issue: what the University could or wanted to offer was modest compared with the offer by the Max Planck Society.

In the discussion with the Rector of the University the message conveyed to me was that I should feel very honoured that the University offered me a Chair. The Rector was apparently unacquainted with the status of the Delft University of Technology (higher than that of the University of Stuttgart) and that I had occupied a well-known Chair there for already 12 years. I felt obliged to inform him that the attraction of Stuttgart for me was established primarily by the Max Planck Institute for Metals Research, not the local University.

Ensuing discussions with the Dean of the Faculty Chemistry, Professor Helmut Bertagnolli, were pleasant and fruitful. Bertagnolli was kind and thoughtful. He perfectly interpreted the situation and tried to comply with my wishes as well as possible.

A major issue concerned both the language used for drafting and the inner structure of a Ph.D. thesis. In Delft all the Ph.D. theses prepared under my supervision were written in English and constituted of a number of manuscripts, already published or submitted for publication (i.e. a "cumulative" thesis, as opposed to a "monographic" one). This choice of language and of format had the following background. Firstly, the lingua franca of science nowadays is English, not Dutch (Ph.D. theses with Burgers as supervisor were usually written in Dutch) and even not German. Secondly, extracting and drafting papers from a thesis, after the thesis had been defended and the former Ph.D. student had left the University, could be a very cumbersome affair putting an extra burden on the supervisor and his co-workers. This could and had led to significant work not being published at all. A thesis composed of a number of manuscripts (say, four) has the enormous advantage that the writing assignment is done only once, i.e. at the same time for a paper and a chapter in the thesis, and, in particular, while the research program runs. Further, the student learns paper writing already in an early stage of the Ph.D. project and feels from the start the pressure caused by the demand to publish the work.

This was not the custom in Stuttgart and I knew it. I would go that far, that I would decide against Stuttgart if one would not allow me to adopt my approach to the Ph.D. thesis also in Stuttgart. I told Bertagnolli accordingly. After our deliberations he said he would fully support me and I should go ahead as desired. However, I should more or less stay silent about our agreement. This prudence derived from the presence of a non-negligible conservative fraction in the professorate of the Faculty. An argument against my approach, that I was well familiar with, said that the Ph.D. student in a "cumulative" thesis would not write the thesis as an individual effort, since scientific papers (the chapters in the thesis) usually have more than one author. My response always was, that presently only rarely a scientific work is the result of one individual and that it should be not only allright but a prerequisite that all who contributed to the research in an essential way should also participate in the paper writing. Of course, the first author, usually the Ph.D. student, should perform the major part of that task, e.g. by writing the first draft of the paper.

Bertagnolli wanted to avoid a corresponding discussion within the Faculty (in fact a major aim of his action as Dean was to avoid conflict and establish harmony; not always a wise choice, but I admit, it worked well as long as he was Dean).[1] He may have speculated that theses from my department and structured as I wished, for a long time would pass unnoticed and uncommented as only the few members of the examination committee (chosen by me...) of a thesis would be confronted with its contents. I accepted Bertagnolli's promise of support and his proposal. Indeed, for the many years ahead with Bertagnolli as Dean, and also thereafter, no such obstacle was put in my

---

[1] In my first year in Stuttgart, Bertagnolli a couple of times urged me stringently to present an Inaugural Lecture to the Faculty/University. I stubbornly refused repeatedly, as I had just given a Farewell Lecture in Delft, where 12 years earlier I had already given an Inaugural Lecture on the occasion of my appointment as Professor in Delft. I couldn't see the sense and certainly lacked any enthusiasm for an(other) Inaugural Lecture. After some time Bertagnolli gave up. In my Farewell Lecture in Stuttgart, 18 years later in 2016 (see Chap. 14), with Bertagnolli, in the meantime retired, present in the audience, I referred to this episode and remarked smilingly that this Farewell Lecture could be considered as redemption for the not given Inaugural Lecture in 1998 as well. During the reception afterwards Bertagnolli said to me that he had no clear memory of our friendly "skirmish" of so many years ago.

way. I even noticed that "cumulative" theses became more common in the course of years, in also Stuttgart.

Some considerable time with this period of negotiations passed, also because some complications emerged, as described in Chap. 10 and also connected with matters as pension arrangements (after my retirement my income would be based on various sources both in The Netherlands and in Germany), before I finally accepted the position in Stuttgart.

Whereas at the University one can speak of a successor on a Chair, this description generally does not hold for the next Director on a Director position that has become vacant in a Max Planck Institute: the next Director can be active on a completely different field than his/her predecessor, as the Max Planck Society looks for selection of (one of) the most eminent scientist(s) labouring a promising field; i.e. usually an abrupt and pronounced discontinuity in scientific direction occurs upon appointment of a new Director. The Director position and the Chair which I was to take over were formerly occupied by Professor Bruno Predel.

Professor Predel

Bruno Predel was well known for his work on the determination of phase diagrams[2] a field of research often described in the German literature as "Konstitution" (i.e. the description of heterogeneous equilibria as presented in phase diagrams). He had started his scientific career at the University of Münster with a Ph.D project devoted to the thermodynamic activities of liquid alloys and the constitution of binary systems. Predel remained true to this field of science along his entire career. He showed remarkable scientific talents which eventually led to his nomination as Chair Holder at the same University in 1970.

Right from the start Predel developed an enormous drive and great productivity in his research. In Münster one may have asked how Predel managed to realize that: he must have worked day and night. Thus the basis was laid for the at long last total number of more than 500 publications of which Predel is author or co-author at the end of his career.

In 1973 Predel was appointed in Stuttgart as Chair Holder at the University and as Director at the MPI-MF. He took 14 of his co-workers with

---

[2] Phase diagrams present fields of stability (i.e. states of equilibrium) for material phases as function of so-called "intensive" state variables. "Intensive" state variables are variables which do *not* depend on the size of the material system: for example, pressure, temperature and composition. "Extensive" state variables do depend on the size of the material system: for example: energy, mass and volume.

him to Stuttgart, in order that, and these are his own words: *"the scientific research could be resumed immediately"* (= damit *"die wissenschaftlichen Arbeiten unverzueglich fortgesetzt werden konnten")*. The accent should be on the adverb "immediately" (= "unverzüglich"): this well fits to what is said above about the Münster time. Also in personal discussion with me, Professor Predel was visibly proud about the performance of his department. Such a smooth transition from Delft to Stuttgart would not be granted to me: I came to Stuttgart alone, which would have consequences.

Unfortunately in the last years of his professional career Predel suffered from Parkinson's disease, with gradually increasing symptoms. He could no longer act as he desired. With greatest struggle and supported by his family and his coworkers at the Institute, he achieved to complete a series of Landolt-Börnstein handbooks (more than 10) on the constitution (see above) of binary alloys. This mammoth work was concluded just before my formal begin in Stuttgart, because Predel in no way wanted to consume time of the coworkers of the department in the presence of an active successor. This scruple is more than just remarkable; I will return to that character trait.

Beyond the class of experts in the field of science of Predel, he is especially well known for his book „Heterogene Gleichgewichte" (= *"Heterogeneous Equilibria"*; Steinkopff Verlag, Darmstadt, Germany, 1982). Much later, Predel was 76 and in bad health, he yet succeeded in completing an English translation, updated with the help of Michael Hoch and Monte Pool as co-authors (*"Phase Diagrams and Heterogeneous Equilibria"*, Springer, Berlin-Heidelberg, Germany, 2004). This book fills a gap in the English scientific literature. I mention it here also because it allows a transition to some more personal experiences.

My contact with Predel

We met each other for the first time during my first visit to Stuttgart described above. He generously informed me then and in the following months about everything I had to know about the Chair and the Institute. His personality can only be described as unassuming and sympathetic, accentuated by perfect manners. He may have lacked a certain degree of crudeness, which, as indicated by a few anecdotes that I know, must have made life not always easy for him at the institute (Fig. 13.1).

Predel's office was in the "Altbau" of the Institute (see below under "*The Max Planck Institute for Metals Research (MPI-MF)*"); it was the former office of Professor Werner Köster, the (re)founding Director of the Institute in 1934

Fig. 13.1   Prof. Bruno Predel (photograph from *"Jahresbericht 2010 der Max-Planck-Gesellschaft"*, *Beilage-1*)

and also occupant of the same Chair at the University that Predel and then I occupied, so to speak as Köster successors. Köster was the permanent Director of the Institute till his retirement in 1965 and still more or less omnipresent until his death at the age of 93 in 1989. The last of about 160 publications with his name as author or co-author (according to the WoS) appeared in the same year... He had definitely put a stamp on materials science in Germany during his active years, especially because of his more administrative assignments. Köster's name was spoken with certain awe. This may make understandable that I was informed, in this sense, that the table and lamp above it in Predel's office were still those of Köster... I was less impressed, as the aura associated with Köster's name was not in proportion with his scientific achievements.

Shortly after publication of the above mentioned English translation of his book, Predel kindly gave me a copy of the book in May 2005. He wrote a dedication, which I represent translated into English here:

Dear colleague Mittemeijer,

This book, that comprises an essential part of our run out field of science, be handed over to you in friendly closeness.

Yours Bruno Predel.

The issue of interest here is the used designation "our run out field of science" (= "unseres ausgelaufenen Fachgebietes"). One could interpret that phrase such that thereby certain sadness is revealed, that the research of his department was not continued by his successor. In no way Predel ever would have exposed his deepest feelings on that matter, and certainly not towards me. He was the first to make the way free for his successor and to give all support if desired explicitly. Only on this single occasion, after I had been active in Stuttgart already for seven years, in this dedication: "run out field of science".… That had hurt him, although in full understanding for the present. It was fully clear to me that I would have to make the same experience upon retiring.

There is perhaps nothing more difficult than to leave an elevated position that is highly respected in society. Manifold stories can be recounted about the "irritations" between successors and predecessors, also regarding Chair Holders and Directors at research institutes. From the very beginning of my time in Stuttgart, nothing of the kind I was about to experience. Predel not only helped me in any way he could, but he literally went "out of the way": his statement "give me an office far away from my successor" is legendary. He really meant that. Then it is all the more remarkable, that, caused by construction work, we initially shared an office, namely *his* office in the "Altbau", Seestrasse 75. It never gave rise to any problem between us. Predel only used the office when I was not in Stuttgart. One evening I was working in this office, sitting behind his writing desk. Suddenly the door opened. Predel was standing there. Evidently it was extremely painful to him. He had thought that I would have been absent, but clearly something had gone wrong with the communication. Also I had no good feelings; after all, it was *his* office and it would have been no problem for me to go elsewhere, where I could also work. So I left, as the younger one, leaving the place to my much older colleague. But Predel could not accept that: he also left. On this evening his office remained unoccupied…

His role as Professor at the University was sacrosanct to Predel; I share that attitude with him. The experiences I lived through with him as examiner, for example in examination committees of Ph.D. candidates, were distinctive. I have rarely met a colleague who strained him/herself to such great extent to calm down anxious candidates. Thus he carried at examinations a very special tie. On this tie an important, very well-known equation from thermodynamics was embroidered: the so-called Gibbs–Helmholtz equation: $G = H - TS$. If the candidate was confronted with a question requiring input of this equation, Predel, kindly smiling, pointed at the equation on his tie, in order to steer the candidate in the correct direction. Or he told an anecdote or small joke to loosen up the atmosphere and to reassure.

The commitment of Predel to students went even much further. Once a student was incapable to perform well during an oral examination; he was simply much too nervous. Predel then made a walk with him, around the building in the Seestrasse, and meanwhile talked with him about the field of science subject of the examination. At the end of the stroll he announced to the student's great surprise that he had passed successfully....

The burden of his illness demanded ever larger tolls. Nearing the end, an entire leg had to be amputated. My wife and I visited him in the hospital. He was in very low mood and we couldn't cheer him up. In 2007 Predel died.

I was asked to deliver a speech at the funeral of Professor Predel on behalf of the Max Planck Institute and the Faculty of the University. I present here the concluding part of that speech, translated into English.

This is the second time that I may speak on the occasion of the passing away of a famous predecessor. Back then, in The Netherlands, I participated in the funeral service for Professor Burgers and now, in Stuttgart, I am standing next to the coffin with Professor Predel. A few parallels come unstoppable to the fore. Both were successful scientists. Both were talented academic teachers. Both were remarkable personalities. Both suffered from pronounced physical disabilities; especially the peculiar feature that both of them (eventually) had only one leg. Whereas I may have learned in my younger years from Professor Burgers how to act passionately and successfully as a scientist, many years later the relatively few but intensive contacts with Professor Predel have made clear that, at long last, acting humanely during life is of all overriding importance. This leads to a last example.

My wife has been for years the secretary of Professor Predel. Often, at the end of a working week, before Predel left the Institute on Fridays, he gave her a small bunch of flowers with the words: "For everything you did for me during this week and at each and every day". His kindness, which came from the heart, was touching; he was a real gentleman.

What this is all about, is, that for many of us, the significance of Predel, as human being on this earth, was at least as important as his impact as scientist.

The Max Planck Institute for Metals Research (MPI-MF)

The MPI-MF has a long history. It was founded in 1921 as the "Kaiser-Wilhelm-Institut für Metallforschung" (= Emperor Wilhelm Institute for

Metals Research; KWI-MF) with Emil Heyn as founding Director.[3] Heyn had also played a major role in launching the "*Deutsche Gesellschaft für Metal-lkunde*" (= German Society for Metals Science and Engineering; DGM), a little earlier, in 1919.

The institute was first housed in rented spaces in Neubabelsberg, close to Berlin. After the death of Heyn in already 1922, the institute was moved to the Staatliche Materialprüfungamt (State Material Testing Institute) in Berlin-Dahlem. There the institute rose to prominence with scientists of names we still know today, as Max Hansen (first compilation of binary phase diagrams), Georg Sachs (plastic deformation), Erich Schmid (Schmid factor/law) and Günter Wassermann (texture (analysis)).

In the first part of the thirties, a time of economic downturn (the Great Depression), the institute suffered from severe loss of financial means, until then provided largely by the metals industry. Moreover, leading scientists left the institute. In 1933 the institute was closed.

Already at the time of closure of the institute in Berlin-Dahlem efforts were underway to reopen the institute at another location, namely Stuttgart. Not only the concentration of metal industry in the southern part of Germany played a role here: the desire of the Land Württemberg to possess a KWI as well as the presence of high calibre scientists at the University of Stuttgart, who would become Directors at the institute, were of cardinal importance. The KWI-MF was reopened in 1934 in Stuttgart (Fig. 13.2a shows the new building opened in 1935, commonly called "Altbau").

The above paragraph represents the history of the move to Stuttgart as offered by the MPI-MF until at least the beginning of the twenty-first century. The reader may have realized that in Germany Hitler took power in 1933 ("Machtergreifung"); Germany had become the "Nazi state". It goes without saying that advanced metals research was of great interest to the war machine that was about to be unleashed in Germany. The change in the political system played a pronounced role in the realization of the reopening/refounding of the institute in Stuttgart. Leading representatives of the KWG, including its President, Max Planck, used the argument of importance of the research of the KWI-MF for the development of armoury technology and thus its war worthiness in general. Indeed the KWI-MF

---

[3] Emil Heyn is a scientist known to me and others because he was one of the first to recognize that plastic deformation of a crystalline solid material can induce the emergence of locally strongly varying, internal stresses as a consequence of the intrinsic elastic and plastic anisotropy of the constituent crystals (also called "grains", cf. footnote 12 in Chap. 8) of a solid material. Such stresses have sometimes been called "Heyn stresses". Their measurement and analysis, by modern diffraction methods, is an important topic of also nowadays research devoted to the indicated "grain interaction" (e.g. see the book "*Fundamentals of Materials Science*", 2nd Edition, Springer, Berlin-Heidelberg, 2021, pp. 350–352).

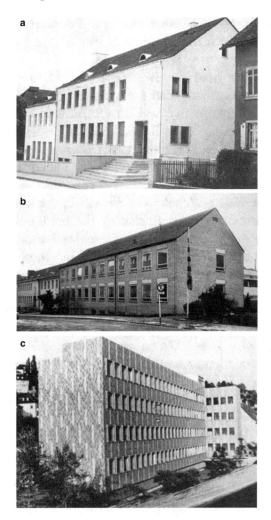

Fig. 13.2 The "Seestrasse complex" of buildings forming the original Max Planck Institute of Metals Research in Stuttgart (source for this figure a–c: https://www.uni-stuttgart.de/universitaet/profil/historie/campus/stationen/stadtm itte/info/info_station_w.html). **a** The first building, usually called the "Altbau" (Seestrasse 75), opened in 1935 in the presence of Max Planck (see main text) (at the time the institute was still called Kaiser Wilhelm Institute for Metals Research (KWI-MF)). **b** From back to front as shown in the picture: the "Altbau" (1935, rebuilt after the Second World War, 1947). "Anbau" (1939; rebuilt after the Second World War, 1949) and the "Erweiterungsbau" (1956), constituting one closed track of buildings along one side of the Seestrasse. **c** A further extension: a new building (1959) at Seestrasse 92, i.e. at the opposite site of the Seestrasse, facing the buildings shown in **b**. The first and third floors were the core of my department during my first years in Stuttgart

would, for example, play a distinct role in the development of new alloys to replace those for which the constituents would be(come) scarce while in war.

That we know this now, we owe Hubert Markl, the President of the Max Planck Society. Germany, half a century after the Second World War, was investigating its role in that war. In 1997 Markl installed a commission to examine the acting of the Kaiser-Wilhelm-Gesellschaft (transformed into the Max-Planck-Gesellschaft in 1948) during the time of the "NS (Nationalsozialismus)", i.e. from 1933 till 1945. The for the general public most well-known result of this investigation concerns the exposition of the involvement of the Kaiser-Wilhelm-Gesellschaft in the medical experiments performed on prisoned humans (including the notorious experiments on twins) by Mengele c.s. Markl, on behalf of the Max Planck Society, of course no perpetrator himself, and in the presence of surviving victims, sincerely apologized publicly in 2001. Markl paved the way for the MPS to accept responsibility for the past.

Also the role of the KWI-MF was investigated explicitly, which led to a booklet by Helmut Maier ("Wehrhaftmachung" und "Kriegswichtigkeit", published as "Ergebnisse 5" in 2002, on behalf of the mentioned committee). The corresponding information presented in the following paragraphs is based on it.

Various documents demonstrate that during the Second World War the KWI-MF was deeply involved in research of importance to the war effort. Here alloy development and nondestructive material testing can be mentioned especially. One looked for substitutes for copper (Cu). Alloys based on aluminium (Al), magnesium (Mg) and zinc (Zn) were focused on. The Managing Director of the KWI-MF, Werner Köster, initiated contacts with high officials of the Ministry of Defense (Rüstungsministerium), the Ministry of Aviation (Luftfahrtministerium) and the Navy (Kriegsmarine) to advertise a Zn-rich alloy developed by the Institute. A series of instruments for nondestructive testing was developed, which were aimed for and found application in arms production, for example to test cannon and gun barrels, to realize aircraft tracking and to search explosive devices, etc. Köster presented in 1943 such results also in a lecture for Albert Speer (Reichsminister fuer Bewaffnung und Munition (= Minister for Armaments and Ammunition), since 1942) and Karl Dönitz (Oberbefehlshaber der Kriegsmarine (= Commander in Chief of the Navy), since 1943).

After the war the allies tried to attract eminent scientists from Germany for work in their own research institutions. Also the KWI-MF was "screened" against this background. Köster, as one of the selected scientists of the KWI-MF, was approached by the French and the Americans, but resisted their offers in order, as he later said, to get the Institute back on its feet again. The Institute was bombed during the war and had moved to locations outside Stuttgart. More or less immediately after the war, efforts were made to rebuild the Institute involving reconstruction of the destroyed Stuttgart buildings (comprising the "Altbau" (Fig. 13.2a) and an "Anbau" from 1939; see Fig. 13.2b). After the transformation of the Kaiser-Wilhelm-Gesellschaft into the Max-Planck-Gesellschaft in 1948, the KWI-MF was renamed, at 1-1-1949, as the Max-Planck-Institut für Metallforschung (MPI-MF).

Köster was arrested in 1945 by the occupying powers. He was released in 1946, wanted to manage the rebuilding of the institute, but was suspended by the occupying powers. In a following so-called "Entnazifizierungsverfahren" (= denazification process) he was categorized as "Mitläufer" (= follower (hanger-on)) and was convicted to pay a sum of 1000 RM (RM = Reichsmark). Immediately after completion of the "Entnazifizierungsverfahren" Köster retook his position at the institute as its Managing Director and participated in the founding meeting of the Max Planck Society in 1948.

The picture one gets of Köster, the longest serving Managing Director of the MPI-MF (from 1934 till 1965), cannot be white or black; it is grey. He became member of the NSDAP (the Nazi party) in 1940, at a time where the horrid, human despising and criminal aspects of the Nazi state must have been clear to anybody willing to see. He appears to have advocated and driven forward the war related research of the KWI-MF. On the other hand, he used his high position to protect co-workers at the institute by claiming that they were indispensable for the performed war related research. In this way he protected people who were endangered because of a political background or a Jewish background (e.g. see the testimony in *N. Becker and K. Nagel, Verfolgung und Entrechtung an der Technische Hochschule Stuttgart während der NS-Zeit (= Persecution and Disenfranchisement at the Technical University Stuttgart), Belser, Stuttgart, 2017, p. 441).*[4,5]

---

[4] This referred to book can be seen as the equivalent of an earlier book published by the Max Planck Society: *R. Rürup, Schicksale und Karrieren (= Fates and Careers), Wallstein, Göttingen, 2008,* devoted to the scientists of the Kaiser-Wilhelm-Gesellschaft (after the Second World War: Max-Planck-Gesellschaft) expelled by Germany under Nazi reign. We tend to think of very famous names as Albert Einstein and Lise Meitner, as examples of those scientists driven out of Germany (in case of Meitner it even was a genuine flight), but already only the thickness of both books, moreover recognizing that these books represent only those expelled from the University of Stuttgart and the Kaiser-Wilhelm-Gesellschaft, testifies of the enormous extent of the process of driving out scientists because of political and racist reasons in Nazi Germany: very many, very good and also less good scientists and students had to move and in many cases their careers were to be aborted for ever.

[5] As a pendant to the conduct of Werner Köster, the morally impeccable and, at a number of occasions, also already immediately after the "Machtergreifung" at the 30th January 1933, courageous

The MPI-MF may have felt embarrassed by its war past. A somewhat amusing, but at the same time revealing anecdote is the following one. In a booklet published by the Institute in 1949, "*25 Jahre Kaiser Wilhelm-Institut für Metallforschung, 1921–1946, Dr. Riederer Verlag, Stuttgart, 1949*", at page 28, a photograph of the official inauguration of the new building of the Institute in Stuttgart at 24-6-1935 has been taken up. We see Max Planck, the President of the KWG, who is handed over by Professor Köster, the Managing Director of the Institute, a goblet with wine from the year Max Planck got the Nobel prize (1918!), as honorary drink. They stand at the entrance on top of the staircase of the "Altbau" shown in Fig. 13.2a. To the left of Max Planck nobody stands. However, considering the photo at plate 8 immediately preceding page 73 of an earlier booklet published by the Institute in 1936 "*25 Jahre Kaiser Wilhelm-Gesellschaft zur Förderung der Wissenschaften, erster Band, Handbuch, Springer, Berlin, 1936*", it is evident that next to Planck at his left side another person was standing, who, to judge from his uniform, is an official of the Nazi state undoubtedly invited to the celebration as an important personality as well. This man clearly had been removed from the photograph in the first mentioned booklet published *after* the Second World War.

Shortly after the booklet on the role of the KWI-MF before and during the Second World War was published in 2002, a meeting was organized, by the acting Managing Director of the MPI-MF on behalf of the Board of Directors, in the large lecture hall of the institute (then already after the move to Büsnau (see below)). I don't remember if the main contents of the booklet were presented by the author of the booklet and/or by one or more of the members of the commission installed by Markl to investigate the role of the KWG in the NS time (see above). After the presentation discussion was

behaviour by another scientist of "Stuttgart", Paul Peter Ewald, can be considered. Ewald was married to a Jewish woman and also his four children were considered as being Jewish. He was a world famous scientist (one of the originators of the dynamical theory of (X-ray) diffraction; also see Chap. 5). Ewald was Chair Holder at the University of Stuttgart (since 1922), with an office more or less opposite to mine (but 70 years later) at the other side of the Seestrasse: a memorial plaque next to the entrance to his office and department refers to Ewald. He was Rector at the University from May 1932 till April 1933. After the "Machtergreifung" at the end of January 1933, in April 1933 at the Conference of the Rectors of the Universities of Germany in Wiesbaden, Ewald observed that his colleagues remained passive in view of the forthcoming, politically and racist motivated dismissals of colleagues. As a response, upon his return to Stuttgart he resigned from his assignment as Rector. In a later meeting at the University, in 1936, with compulsory attendance of all Professors and Lecturers, where a letter from the Minister of Science (= "Wissenschaft"), Education (= "Erziehung") and National Education (= "Volksbildung") was read, that centered around so-called "objective science", Ewald became such appalled that he stood up and left the meeting hall. This was possibly the drop that made the bucket spill: soon thereafter, in the beginning of 1937, he was called to a meeting with the acting Rector and forced to request his dismissal. In the same year Ewald emigrated to England (see the referred to book in the main text).

possible. The public remark made by one of my older colleagues is imprinted on my mind. He said that the picture sketched in the booklet was too negative and that those leading the institute at the time could not have acted differently. A deep silence of embarrassment of the audience in the lecture hall followed.

After the Second World War, the destroyed Institute buildings were rebuilt: the so-called "Altbau" (from 1934) and the "Anbau" (from 1939) were usable again in 1947 and 1949, respectively. After the renaming/refounding of the Institute as Max-Planck-Institut für Metallforschung at 1-1-1949, the Institute began to flourish again and regained its worldwide recognition. Through the years it also expanded its manpower and research fields as also illustrated by the construction of additional buildings. Thus the so-called "Erweiterungsbau" became available in 1956. "Altbau", "Anbau" and "Erweiterungsbau" constituted one closed track of buildings along one side of the Seestrasse (Fig. 13.2b). At the opposite side of the Seestrasse, in 1959 a new building, with the address Seestrasse 92, became ready to move in (Fig. 13.2c). Finally, a further "Erweiterungsbau", behind the "Erweiterungsbau" of 1956 was built and completed in 1968. This describes the Seestrasse "complex" of the Institute at the time of my appointment in 1997, which would also become the (initial) location for my department (see below under "*My Department; Early Experiences*").

The Institute also comprised the Powder Metallurgy Laboratory completed in 1968 on a piece of ground in Büsnau, a suburb of Stuttgart, where later, in 1975, also the Institute for Metal Physics as part of MPI-MF (see above under "*Start in Stuttgart*"), together with the sister Max Planck Institute for Solid State Research, occupied a new building.

This dispersed nature of the departments of the MPI-MF certainly was not a desirable arrangement. Eventually, next to and connected with the building built in Büsnau for the "Institute for Metal Physics" of the MPI-MF and the Max Planck Institute for Solid State Research, a new building was completed in 2002 where those departments of the MPI-MF until then contained in the "Seestrasse "complex" (i.e. including my department) could move in. Thus a Max Planck Stuttgart campus had been realized where two large Max Planck Institutes had become close neighbours, presenting an enormous research power at one location (Figs. 13.3 and 13.4).

The MPI-MF flourished also in this new arrangement. Its high standing can for example be illustrated by its impressive citation record as exhibited by published citation rankings of research institutes worldwide. As in the past, its research was of highly dynamic character. In the course of years, three factors led to an important change in research direction.

Fig. 13.3 The appearance (aerial shot) of the Max Planck Stuttgart Campus (in Büsnau, a suburb of Stuttgart). The MPI for Metals Research (after 2011 MPI for Intelligent Systems) and the MPI for Solid State Research as neighbours at the same site. The MPI for Solid State Research is concentrated in the lower, right part; the MPI for Metals Research is located at the higher, left part. The building appearing light grey in the photograph (cf. Fig. 13.4) was completed in 2002 and housed my department at the first floor and part of the second floor above ground level (photograph provided by the Max Planck Society)

Fig. 13.4 The new building of the Max Planck Institute for Metals Research opened in 2002 (cf. Figure 13.3) (Source *Max-Planck-Institut für Metallforschung, Stuttgart, Bauten der Max-Planck-Gesellschaft, Bauabteilung, Max-Planck-Gesellschaft, München, 2002; photographer: H.G. Esch*)

The field of materials science has widened pronouncedly and this has led in the MPI-MF to research on, for example, ceramic materials, bio(mimetic)materials and nanomaterials, thereby opening exciting and challenging new research fields as diverse as thin film behaviour, interface chemistry/physics, biophysics and even synthetic biology. Evidently, the name of the Institute covered its research area only partly. Thereby for the "world outside", also within the MPS, an anomaly became apparent and voices became loud to restructure the Institute. The MPS, always keen to identify new, highly promising research fields, developed the desire to initiate a focus on robotics/intelligent systems/artificial intelligence. At about the same time three of the Directors of the MPI-MF left the institute (retirement or appointment elsewhere) and divergence of opinion among the acting Directors of the MPI-MF emerged regarding the further development of the field. This situation was concluded by the Presidency of the MPS by deciding that the Institute should develop in the direction of "intelligent systems", i.e. still with a connection to materials science, but this field of science will not be a dominant characteristic. Thus the retiring or leaving Directors of the MPI-MF will be replaced by "intelligent systems" scientists. Consequently, a gradual, taking years, change of the Institute took and takes place (at the moment of writing this text, in 2021, still a couple of the "old" Directors are active within the Institute). It then was only fitting that in 2011 the Institute was renamed as Max Planck Institute for Intelligent Systems. I retired in 2016.

My Department; Early Experiences

At the start of my work in Stuttgart, the spaces available to my department were more or less as dispersed as the above described accommodations available to the Institute: the core of my department consisted of two entire floors of the building completed in 1959 (Seestrasse 92; Fig. 13.2c) opposite to the closed track of the buildings "Altbau", "Anbau" and Erweiterungsbau of 1956 (Fig. 13.2b); here also the X-ray diffraction (diffractometry) and calorimetric apparatus were located. My office was on the first floor, practically above the main entrance of this building. Two smaller parts of my department were located in the older buildings on the opposite side of the street (here X-ray diffraction apparatus for photographic techniques were available) and in the "Erweiterungsbau" of 1968 (where the equipment for scanning Auger and X-ray photoelectron spectroscopy was installed). Because I occupied a Chair at the University as well, some university personal and university space also belonged to my department: a floor of the university building in the Wiederholdstrasse (a cross street of the Seestrasse) was assigned to my Chair.

In view of its scattered geometry, I worried somewhat that it might be difficult to establish significant cohesion in and control of the department. This concern appeared unjustified. Apart from a joint, daily coffea break, weekly meetings were held with the entire department, where one (Ph.D.) student or scientist of the department gave a presentation with following discussion. My main instrument to keep tight control of the entire department was a meeting, at least once per month, in my office separately with each (Ph.D.) student or postdoc, together with the senior scientist responsible for the daily supervision of the Ph.D. student or post doc, with intensive discussion on the basis of a written report about the progress in the past month. Discussing of drafts of papers usually required additional meetings. I maintained this approach, in fact a copy of my working method in Delft, till my retirement. It proved to be a highly effective procedure to generate scientific success and guarantee high productivity of the department.

When I was appointed in Delft as Professor, I was to lead a group that I knew very well: I had done my Ph.D project there and thereafter I had maintained cooperation with this group. Upon becoming their Professor and my return to the group, I did not plan a discontinuous change of research direction for that group. Starting in Stuttgart would become a totally new experience for me.

I came to Stuttgart alone and was confronted with a large group of co-workers and technicians that I was unfamiliar with. Additionally I had to fill in a number of vacancies. Particularly a more or less abrupt change of research direction for the existing, "inherited" scientists would have to occur. I knew what I wanted to do: to develop a department focusing on phase transformations in solids, bulk- and nano-materials, with emphasis on the thermodynamics and kinetics of these processes.

After my first discussions with the "inherited" scientists I was prepared to meet some initial resistance of about three of the senior scientists that I had taken over. I was willing to accept some transition time, where a running project from the past was gradually concluded while at the same time a project initiated by me should progressively become running. In one case the scientist involved fully accepted the situation and in the end became a pleasant co-worker: together we largely focused on the analysis of the kinetics of phase transformations and became very successful; I keep very pleasant memories from my many years of cooperation with Professor Sommer.

In case of another "inherited" co-worker, of already advanced age, I was less successful. Apparently, in the past he was used to a situation where he could act completely according to his own liking. After I had been active in Stuttgart for a number of months, I once entered his office, without advance notice, to be confronted with three guests from overseas invited by him. I was informed that these guests were regular invitees staying each year a number of weeks during summer time at the institute. I was expected to provide the travel and accommodation finances for these visitors. My co-worker did not seem to understand the situation: I was not there to support him by continuation of what he had done until my arrival, but he was there to support me to realize my research plans. Our relationship remained manageable, but he felt frustrated and went with early retirement after some years. I did not regret his leave, but overall this was an unfortunate development: he was a good scientist and, after our difficult start, we did publish together some good work.

I had tried to take two young promising scientists of my group in Delft with me to Stuttgart, but they declined. Yet, through the years, my presence in Stuttgart apparently exerted an appeal on some Delft students and former Delft Ph.D. students. Firstly, at my start in Stuttgart in 1997/1998, one of my talented Master students applied for a Ph.D. project with me in Stuttgart, that I was glad to offer him. Later, two further students, with completed Master theses from Delft, joined me in Stuttgart and completed their Ph.D. projects in 2005 and 2013 respectively. Two of my former Ph.D. students in Delft, Peter Graat, for a few years, and Lars Jeurgens, for quite a number of years, accepted, respectively, post-doctoral and senior scientist positions in my department and became successful, particularly Lars Jeurgens. In view of my initially failed efforts to take a few scientists from my Delft group to Stuttgart, the unplanned establishment of a small Dutch presence in my Stuttgart department, almost up till my retirement, was an unexpected but pleasant phenomenon: these co-workers contributed at high level to our science.

Through the years a large number of post-docs and Ph.D. students from many foreign countries, especially China and India, were part of the department. Already only against this background the language for daily communication was English. This formed some obstacle for some of the technicians in the department. I had decided that also all technicians and the secretary should take part in our weekly departmental meetings: even, if because of the language used (English) and/or the scientific contents of the lecture and discussion, it would not be possible for all of the attendants to understand all what was said, one yet got an impression of the person lecturing that could only be helpful for interaction in the daily work. And, of course, also daily matters beyond science, for example dealing with institutional affairs, were discussed (mostly by me), but then bilingually, so that it was assured that everybody understood.

Through the years, the number of Ph.D. students in my department varied from usually about 10 to rarely 14. In my view, and with my method of close supervision described above, already only because of the limited amount of energy available to a single man, a larger number of Ph.D. students cannot be managed such that the supervisor can claim to be scientifically active on front level him/herself. The department head, Director, then becomes more a science manager; the supervision of the Ph.D. students and postdocs is then fully in the hands of the co-workers of the Director.

I have witnessed departments in Max Planck Institutes where about 30 Ph.D. students were employed. A Ph.D. student from such a department, having heard from my Ph.D. students of my completely different approach, once spoke with me and complained he had seen and spoken personally with his doctor father about his project at the moment of his employment and would have a personal discussion about his work with him a second time at the end of his Ph.D. project term. This may be a bit exaggerated, but not too much. In my later years in Stuttgart, and after such complaints from various sides, the Max Planck Society formulated rules to avoid such excrescences.

The main reason for me to come to Stuttgart was the offer of personnel and financial means large enough not to be necessitated to apply for external funds. Only in cases where industries and university groups from outside approached me, for cooperation in a project that was attractive to me, I decided to participate and together we applied for corresponding finances. The principal reason for participation then was my interest in the project, much less the position/means to be obtained. Thus I had scientifically rewarding cooperations with various departments of Bosch and Honda and enjoyed taking part with colleagues abroad in a few projects financed by the European Union. This never exceeded more than the acquisition of one or two Ph.D students or postdoc positions. On this basis my department never comprised more than on average about 45 people. My approach, as indicated here, was not the rule. It was clear to me that a different attitude was possible as well and perhaps more "normal". Quite a lot of Max Planck Directors were very active in external fund acquisition and corresponding success was advertised broadly; indeed this activity was a necessity if (group-size) wishes exceeded those the primary MPS fund allowed to realize (see above paragraphs).

For me, coming from a university where I had to apply, more or less to exhaustion, for external funds and was faced with imposed reduction of university funds (described in Chap. 10), the position as Director at the MPI-MF felt as a "luxury", for which I was and remain extremely grateful. No doubt my years as scientist/Director within the Max Planck Society have done much good to my science and have become my most successful and productive time. It took some time for at least some of my colleagues to understand that I was somewhat "different", that I did not strive for more funds and more "importance".

In retrospection two early experiences during meetings of the Board of Directors of the Institute were of pronounced importance:

A substantial part of the finances of the Institute were not assigned to the specific departments but were to be utilized by agreement by the Board of Directors. It became clear to me that, apparently until my arrival at the Institute, part of the finances were distributed among the departments in proportion to their size and that department size incorporated also positions acquired by external means. Consequently a department as mine, where external fund acquisition was not (planned to be) important, would be assigned significantly less money as other departments where external fund acquisition was pronounced. I argued that this was unfair, because it was the personal decision of a Director to enlarge his/her department by applying for external means as it was also the personal decision of a Director not to do that. I had come to Stuttgart precisely for in principle not needing to apply for external funds. These finances provided by the Max Planck Society should therefore be distributed among the departments on the basis of their sizes as determined only by the personnel in the department as defined by the outcome of the negotiations of the Director with the Presidency at the time of his/her employment. My colleagues eventually agreed with my line of reasoning and that was the basis for financing the departments from that moment on.

Looking back, the second experience may have been one of larger consequence for the Institute as a whole. I proposed to change the name of the Institute from MPI for Metals Science to MPI for Materials Science. I argued: such name changes had occurred all over the world (the Laboratory of Metallurgy in Delft had become the Laboratory of Materials Science a few years earlier) and reflected also the changes (broadening) of our field of science. Indeed the MPI for Metals Research was no longer at all only confined to research on metals…. My colleagues were against this proposal. It was said (i) such a change of name would likely not be accepted within the Max Planck Society because there were other MPIs with activity in also the field of materials science and (ii) the name Max Planck Institute for Metals Research was worldwide known and respected; we would damage our recognizability and standing in the world by a change of name. In view of what happened in

2011 (see above under *"The Max Planck Institute for Metals Research (MPI-MF)"*) I still believe that it would have been better to apply a name change as proposed by me in 1998: the historic name did not cover the breadth and novelty of the field covered by the Institute and moreover it certainly made also within the Max Planck Society a dated, old-fashioned impression, as if hot, new areas were no longer identified and laboured by also the newly appointed Directors in the Institute.

Manners and Mores

Upon his arrival in Stuttgart Eric was prepared to be confronted with a society where manners, i.e. the way people deal and communicate with each other, subject to prevailing standards of politeness in daily life, would be a little different from what he was used to in The Netherlands. Yet he would experience some surprises.

In Delft practically nobody spoke to Eric calling him "Professor"; with very few exceptions everybody in his group used his Christian name and "dutzte" him.[6] Strikingly, this generally did not hinder his co-workers, technicians and students to leave him the role as group and research leader and thus to recognize who was the "boss". Eric knew that social relations in Germany would be more formal.

Rather shortly after his start in Stuttgart Eric made the mistake that he invited one or two co-workers of about his own age to use Christian names vice versa and to "dutzen" each other. This could lead to serious misinterpretation: one co-worker apparently thought to be at the same "level" as the superior, which led to misunderstandings in moments of decision taking. Whereas using Christian names and "dutzen" in Delft did not necessarily vapourize the social, hierarchical distance between the Professor and the co-worker, this clearly could easily happen in Stuttgart. Eric learned this lesson and, although against his nature, he therefore kept formal distance from now on, also to his colleagues, which was the best attitude. Yet, after many years of cooperation, using Christian names and "dutzen" each other gradually developed with a number of his close co-workers, but without impairing the

---

[6] Whereas in English both a close friend or close family member and a stranger or somebody at higher social level can be addressed with "you" (so that one cannot guess upon usage of this word how small or large the social distance between the speaker/writer and the addressed one is), in, for example, the Dutch and German languages two words are in use for "you": "du" and "Sie" in German and "jij" and "u" in Dutch. This explains the German verb "dutzen" (in Dutch "tutoyeren") as characterizing people who use "du" or "jij", respectively (close social distance), to address each other.

hierarchical distance, a necessity for managing a large research department. Typically, the Ph.D. students in Stuttgart did not "dutzen" Eric, whereas this was the norm with his former Ph.D. students in Delft.

A more or less reverse observation about societal rules and manners is made upon comparison of the Ph.D. examination procedures in Delft and in Stuttgart, briefly described as follows.

In Delft the entire examination takes place in public in front of a candidate-interrogating committee usually composed of about eight Professors, including the (substitute of) the Rector. All Professors are dressed in their talars, with the typical berets on their heads upon entering and leaving the hall of examination in a row, preceded by the likewise solemnly dressed mace-bearer of the university, while the audience has to stand up. During the examination the Professors sit behind two tables to the left and right of the candidate, all in front of the now sitting audience. The candidate stands during the examination, wearing a tailcoat. The whole ceremony thus is a dignified event that is taken serious by all attendants (although, having come that far, candidates can effectively no longer fail, albeit this is in principle possible; to prevent such embarrassment, the work, on the basis of the thesis, was examined in a previous stage by a smaller committee of Professors closely connected with the topic of the thesis work). Eric appreciated this procedure and ceremony, as it reflects the importance and significance of the moment, concluding a period of 4–5 years of strenuous and successful scientific work by the candidate.

In Stuttgart Eric was informed that talars (and berets) were not in use; this would be an outcome of the "movement/revolution" against traditionalism and authoritarian aspects of society, that had swept through, especially, the universities of Europe in 1968. The Professors in Stuttgart indeed during none formal ceremony of the University, as, for example, the opening of the Academic Year, wore talars… And thus Eric sold his talar in Delft before he left for Stuttgart.

The Ph.D. examination in Stuttgart has an almost casual character. It happened that Eric had to inform the candidate in advance that an appearance in jeans and t-shirt was less appropriate. Unfortunately the Professors involved also not always were dressed in a way that reflected the significance and weight of the moment. The examination takes place in an office under exclusion of the public by a committee of three Professors. Usually a pleasant discussion, sometimes almost derailing to a coffea chat, develops. Also here, failure to pass formally is possible but in practice does not occur.

It is amazing that especially in a significantly less formal, egalitarian country as The Netherlands a ceremony as the official Ph.D. examination is maintained in its formal fashion as described above. I have been member and external examiner in also Ph.D. examination committees in countries as Sweden, Finland, Denmark, the United Kingdom, Belgium and France. In what may be considered the more progressive, liberal, informal and egalitarian societies (the Scandinavian countries and The Netherlands) the more serious, formal and solemn Ph.D. examinations take place.

One reason behind the almost casual character of a Ph.D. examination in Stuttgart can be that, certainly within the field of chemistry, obtainment of a Ph.D. degree is considered as the "normal" conclusion of an academic education and that the time period for a Ph.D. is only 3–4 years. Thereby the Ph.D. examination may regrettably be reduced to a less monumental event in academic life (also see the discussion in Chap. 6).

A general characteristic of a person cannot at all be given on the basis of his/her nationality. Nevertheless it seems possible to describe a people as a whole with positive and negative traits and manners, which, however, can be governed by prejudices. Knowing this, the following is remarked and should be considered with utmost prudence.

The Dutch are known for a liberal attitude.[7] This can be considered a positive property. On the other hand, they are "direct" to clumsy to even rude in their social interaction and more than rarely have bad manners.

Concerning the Germans as a people, Eric had to learn that there exist pronounced attitude/character differences between the North (e.g. Prussia) and the South (e.g. Bavaria, Swabia). Stuttgart is Swabian. Swabia as representing a separate culture had not been on Eric's mind.

Within Germany the Swabians are highly respected for their diligence and their innovative and engineering power (inventor spirit): not for nothing large automotive industries as Daimler, Bosch, Audi and Porsche and many medium-sized companies (= "mittelständische Unternehmen"), who by themselves are world leaders in their fields, are located in Swabia (Land: Baden-Wuerttemberg). Quite a number of the scientists at the Institute had a Swabian origin. This even more had been the case with the former generation of Directors, where a disproportional part (as compared to the size of the German people) had this background. This is no longer the case, also because a substantial fraction of the Directors now is from abroad.

Upon his start at the Institute Eric got from the Works Council a vocabulary Swabian–German. This was meant to be funny. The local government

---

[7] This liberalism has its limits. Whereas the Dutch were forerunners in accepting and providing legal framework for, for example, abortion, formal marriage of homosexual or lesbian couples and self-determined death and they had a liberal drugs policy, in recent years some populist political parties have emerged in The Netherlands with anti-islam and anti-immigration programs that have drawn something like 25% of the votes in recent elections...

advertises the Land Baden-Wuerttemberg with the saying: *"Wir können alles. Ausser Hochdeutsch" (= "We can realize everything. But (can)not (speak) real German")*. Swabians are also typified as being "business savvy" to shrewd, as Eric experienced also at the Institute. However, some special experiences during his life in Stuttgart as a private person did impress him more[8]:

Upon buying a small house in Botnang, a suburb of Stuttgart, Eric made acquaintance with his future neighbours to the left and to the right of the house. These meetings ran in a normal, friendly atmosphere, which did not prepare him for what happened next. After he had moved into his new home and returning at the end of the next day from the Institute, Eric found a letter in his mailbox from one of his neighbours. The letter explained to him that the fence around the garden in front of the house (which consisted of rows of conifers) was too high. In the letter Eric was informed that this height was incompatible with a specific municipal ordinance and he was commanded to cut these conifers down to acceptable height (for sure, the conifers would not survive such drastic cut). Eric was flabbergasted. Why had this neighbour not approached him about this matter during their first meeting? Why a letter, whereas a discussion among neighbours would be the normal way for bringing up the "problem"? Eric's irritation was that large that he, somewhat agitated, immediately went to his neighbour and made clear that this was not the way to handle problem points between neighbours. Eric had lived many years in his own house in Delft and always had maintained pleasant relationships with his neighbours. This was a new and extremely unpleasant experience. Moreover, he had bought the house in this condition and could only conclude that his neighbour should have solved the matter before, with the former owner of the house. Eric left the fence as it was. The relationship with this neighbour remained problematic: the neighbour loved to idling the engine of his car on Saturday's on top of a grid, in the pavement of the entrance to his house, with direct atmospheric access to the sleeping room of Eric's house one floor lower (Eric 's and the neighbour's houses stood on

---

[8] The Swabians like their food specialties (as their tasteful "Maultaschen", consisting of a cover of pasta filled with minced meat, spinach, bread, herbs and more) and their wine. And they are proud of it: On his first evening in Stuttgart (see *"Start in Stuttgart"* at the begin of this chapter) Eric was invited to join a meal in a restaurant with a couple of his future colleagues. One of these was a genuine Swabian. After having selected the wine, he informed Eric that all the wine that was produced in Swabia was consumed by the Swabians themselves. The message of this remark was clear: it should be interpreted as an indication of high quality of the wine, such that the Swabians would prefer to keep the wine for themselves rather than exporting it. Drinking this (red) wine, often a Lemberger or Trollinger or blends thereof, was not a positive experience for Eric; this red wine was much too "light" for his taste and generally did not impress at all. The thought, that a low rather than high quality of the Swabian wine then is the barrier for its exportation, imposes itself. To be fair and balance this critique, later, but rarely, Eric discovered also very good wines from Swabia, for example he enjoyed a delicious Kerner (white wine) from the "Felsengarten" (Neckar Basin) region.

a slope), so that Eric and his wife, Marion, could enjoy inhaling the exhaust gases of neighbour's car during the day they could stay in bed a little longer than during the week.

Quite some years later Eric and Marion lived in a large penthouse in Sonnenberg, another suburb of Stuttgart. The whole house was composed of 5–6 apartments. They had bought the penthouse as the place to stay, i.e. especially also after retirement. This plan was found to be untenable: Through the years the atmosphere in the house deteriorated. One of many peculiar, less important but typical experiences made, was the following. All owners had agreed to put a "no parking" sign next to the gate where one left the plot onto which the house with its garages stood. The sense was to assure free sight for the outgoing cars, so that no collision with the traffic on the street could occur. To Eric's and Marion's utter amazement and irritation they noticed that the family who had been the strongest protagonists for this measure, did park their own car or let park their visitor's car on also just this place where the "no parking" sign had been put. Evidently, rules are only meant for "the others".

Neighbour quarrels may be considered to be of narrow-minded, provincial nature. Yet they can be a burden for daily life. Unavoidably in view of their experiences, Eric and Marion could not escape the thought that this problematic is a characteristic of Swabians, but they accepted statements that the accumulation of such experiences bestowed on them was just by chance. They needed and had decisive help in legal matters from a good friend, a judge of profession. In the end, after ten years in their penthouse, Eric and Marion decided to leave Stuttgart. They now live to satisfaction in Heidelberg.

Initially I had decided that leaving Delft would mean conclusion of my work on nitriding (cf. Chap. 8). The move to Stuttgart I thought to be the moment to make time available for other exciting topics. And so it happened. However, after a couple of years in Stuttgart, I changed my mind, not least stimulated by an experience that is described below: it became clear to me that research on the interaction of metals and gases and the role of interstitially dissolved elements, as nitrogen and carbon, in metals did still imply the revelation of fascinating phenomena not observed before, which allowed a lot to be learned about the behavior of materials.

With the help of a very able technician a nitriding facility was built up. This took a very long time, which annoyed me considerably. The delay was in particular caused by safety measures that had to be realized: for example, whereas in Delft we had let simply escape the used ammonia and hydrogen gases into the atmosphere outside the Laboratory by a tube through the nearby window…., such a "solution" was impossible in Stuttgart. I had to accept the German safety regulations and, of course, I had no real objections; I could not go out of the way of my responsibility.

Over the years the facility was extended such that carburizing and nitro-carburizing were possible as well, by involvement of additional furnaces and additional tracks for gas preparation, cleaning and regulation. The fine point was, different from many other such facilities in the world, that we could control the chemical potentials of the nitrogen and the carbon, both separately and together, which gave us an enormous advantage compared with competitors.

The discovery of amorphous precipitates in a crystalline solid

A scientist may experience moments of genuine discovery only rarely. They occur if an observation is made or a flash of insight occurs that, at the time, is contrary to what is common sense, i.e. contrary to what is "normally" to be expected. I was subjected to such an emotion shortly after I became active in Stuttgart.

In a *crystalline* material the atoms/molecules are arranged in a regular, periodic manner. This contrasts with *amorphous* material where a more or less chaotic, near random atom/molecule arrangement occurs. The crystalline, instead of amorphous modification of a solid compound is the one of lowest energy and thus the one preferred by nature. Hence, if within a crystalline solid that is in a supersaturated state, particles of a new phase start to develop (this is called "precipitation"), then these particles will be crystalline. At least this is the "normal" observation. And this was what I expected in our experiments on the precipitation of silicon nitride in nitrided iron-silicon (Fe–Si) alloy, a project executed together with Mohammad Biglari.

Mohammad Biglari had been a Ph.D. student in my Delft group and had performed a project on the nitriding of iron-alumium (Fe–Al) alloys (also, see Chap. 10).

Mohammad is of Iranian origin and before arriving in The Netherlands already had a more or less adventurous life behind him. He had fled from Iran, via Sweden, to The Netherlands with his wife and young son (this son, Mostafa, about 20 years later, would also do a Ph.D. project under my supervision, then, of course, in my department in Stuttgart). He was a remarkable man only 10 years younger than I was (at the time of his promotion in 1994 he was 34 years old; according to Dutch standards rather "old"). He had decided that he wanted to integrate himself fully in the Dutch society and, after having finished his Master project, he wanted to complete a Ph.D. project at "whatever it takes". To provide his family a minimal state of living, he accepted all kinds of jobs. Especially well I do remember that for a considerable period of time he worked as a baker at night. Yet, he appeared during the day at the Laboratory to perform the experiments for his project. He was modestly talented as a scientist, but, one could say as efficacious compensation, I have seldom met such energy and determination. He was also charming and one could not stay "angry" with him, as I sometimes felt I should. I liked him. Kees Brakman (see Chap. 10), as his daily supervisor, had a strong and good relationship with him and felt similarly as I did about him.

In the crystalline, solid iron-alumium alloy, supersaturated with nitrogen, aluminium-nitride particles precipitate. And, of course (see above), these aluminium-nitride particles are crystalline. We were interested particularly in the kinetics of this process and this was a focus of Mohammad's thesis. Some considerable time after completion of his Ph.D. project, Mohammad proposed to complete unfinished work on silicon–nitride precipitation in iron-silicon (Fe–Si) alloy, supersaturated with nitrogen, which we once had started to investigate the kinetics of that process. We already knew from our first experiments that the precipitation process was unusually, extremely slow and that had been one reason to leave that additional investigation unfinished during his time-limited Ph.D. project. Mohammad now had a fulltime job elsewhere of more management character and I had just started in Stuttgart; this research endeavour thus promised to be a "long game" requiring staying power of both of us. Our meetings, taking place during a number of years, were restricted to evenings where I happened to be in Delft and, after I had picked him up from the railway station, we could discuss the usually modest progress made during a joint evening meal. A main obstacle appeared to be the puzzling nature of the nitride precipitates.

The nitride expected was silicon nitride of the composition Si_3N_4. Because the possible crystal structures of this nitride are known, we looked for reflections originating from this nitride in X-ray diffractograms (cf. *"Diffraction Analysis as a Tool for Determining the Constitution of Matter; Regularities and Irregularities of the Atomic Arrangement"* in Chap. 5). To our amazement we could not find any such indication for the presence of the nitride. We checked and rechecked everything that might have gone wrong in the diffraction experiment and performed a number of control experiments. All to no avail: the nitride remained "invisible". Actually we could have arrived already then at the correct conclusion. However, we were caught in the "world of what we expected", namely a *crystalline* precipitate; we were not able to make the mind jump necessary to arrive at understanding, which is a mental problem now and then experienced by scientists confronted with observations which do not comply with the governing "paradigm".

We performed additional measurements. The mass increase of the specimen upon nitriding was measured. Knowing the amount of silicon in the specimen, the observed mass increase very well agreed with the amount of nitrogen necessarily to be taken up for the nitride Si_3N_4 to precipitate. Then, using a spectroscopic method, we determined the type of chemical bonding of the nitrogen in the specimen. The chemical bonding type was found to be compatible with the presence of all nitrogen as in Si_3N_4.

As a final step we decided to apply high resolution transmission electron microscopy. With this method the (atomic arrangement in) tiniest precipitate particles can be made visible. However, with iron-based specimens, which are strongly magnetic, this method is not easy to apply.[9] After having overcome significant experimental trouble, we finally obtained good micrographs. I will never forget the excitement we experienced upon looking for the first time at the pictures: clearly, the precipitate particles were *amorphous*!

Looking back at this short history, it strikes me still that this unavoidable conclusion was not drawn by us already after only the X-ray diffraction experiments. Instead it had taken us a long time to overcome our "narrowmindness", constrained and biased as we were both by experience and what was considered "normal" in the literature. Thus this story, of a small discovery, illustrates a lot of how science can run.

The precipitation of an amorphous phase in a crystalline solid was highly unusual, counter-intuitive and, to our knowledge, not observed before. So we decided to publish this result more or less immediately, as a short communication, in a well-known journal in the field of materials science (*Scripta Materialia, 41(1999), 625–630*). In that note we also proposed that the stability of the amorphous state of the particle was due to a lower energy for an amorphous(particle)/crystalline(matrix) interface than for a crystalline(particle)/crystalline(matrix) interface, for the same partners joining the interface. Formulated in a general way for precipitation reactions: the relatively low value of the amorphous/crystalline interface energy may overcompensate the larger bulk energy of the amorphous modification for *small enough* particles (because the smaller the particle, the larger the relative

---

[9] Within the (transmission) electron microscope (actually within the "column"; along its axis the electron beam passes that is used for image formation) magnetic fields occur due to the operation of the applied magnetic electron lenses. Interaction with the magnetic field invoked by the magnetic specimen complicates handling of the microscope. The effect can even cause the disintegration of the mechanically very vulnerable, thin, electron transparent specimens(foils), if these are magnetic. Thereby specimen fragments may be attracted and deposited onto the pole shoes of the magnetic lenses…., which is obviously undesired.

contribution of the (energy of the) atoms at the interfaces with the matrix): the very first stage of precipitation then can be development of amorphous instead of crystalline particles.

To my amazement, for a long time this paper apparently remained unnoticed. At least it drew no citations. It took more than 10 years before citations were noticed by the Web of Science (WoS). It is possible that the general meaning of the result was missed and that the paper was considered as of relevance only for research on nitriding.

At about 2010, a Ph.D. student from India, working under my supervision in Stuttgart, Sai Ramudu Meka (called Sairam), had just finished his Ph.D. and would stay for a couple of years as postdoc. We had discussions traversing in a crisscross manner the field of materials science. I must have told him of my result of more than 10 years ago on the emergence of amorphous instead of crystalline precipitates. I presume Sairam may have thought that the result obtained back then by me and Mohammad Biglari was possibly suspect, i.e. erroneous, and that that would be the explanation for the lack of citations until then. Therefore he wanted to check for himself, if the result could be reproduced. Thus, without notifying me, he performed the appropriate experiments. After some time he came to me showing that he indeed also obtained *amorphous* silicon–nitride precipitate particles. Understandably I was glad about that outcome.

This was not the end of the story:

Further experiments on the development of the silicon–nitride particles now were performed. At temperatures higher than the one originally applied by me and Mohammad (650 °C vs. 570–580 °C), surprisingly, a highly peculiar shape for the amorphous silicon–nitride particles was observed: a strangely eight-legged, octapod-shaped morphology for the nanosized, amorphous, silicon–nitride particles (see Fig. 13.5a, b). At the lower temperatures amorphous, cubical-shaped particles developed, as shown by us in the earlier work. Amorphous octapod-shaped precipitate particles had never been observed before. So we published this experimental result, together with a speculation about the origin of this peculiar morphology, as a short note (*Philosophical Magazine Letters, 93(2013), 238–245*). Strikingly, the publication of this paper, together with a companion one published in 2012, may have been the vehicle to at last draw more attention to the original work published in 1999, as suggested by the citation history.

In February 2002 we could eventually move from the Seestrasse to the new building in Büsnau (Figs. 13.3 and 13.4). My department now was much

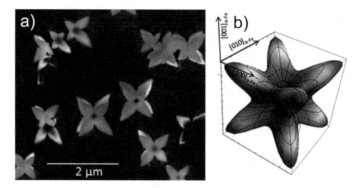

Fig. 13.5 The strangely eight-legged, octapod-shaped, amorphous silicon–nitride (Si_3N_4) particles, developing in iron (Fe) alloyed with silicon (Si), upon nitriding at 650 °C (adapted from *S.R. Meka, E. Bischoff, B. Rheingans and E.J. Mittemeijer, Philosophical Magazine Letters, 93(2013), 238–245*). **a** Scanning electron micrograph. The star-shaped, amorphous silicon–nitride particles show eight legs (1 μm = 10^{-6} m). **b** A schematic illustration of the amorphous silicon–nitride particle shape and its orientation with respect to the surrounding crystalline iron matrix (here the so-called ferrite modification of iron is meant, indicated by α-Fe in the figure; the numbers between the square brackets denote crystallographic directions in the iron)

more concentrated; we occupied the entire third floor (in fact the first floor above ground level) and a part of the fourth floor. This and, no doubt, the virtue of a brand-new building and brand-new office and support facilities were a pleasure and an improvement as compared to the situation in the buildings constituting the now former Seestrasse complex of the Institute. Everything in Büsnau was large; the two Max Planck Institutes together, the one for Metals Research and the one for Solid State Research, represented an impressive concentration of people (about 800 scientists, technicians and administrative personnel were working there) devoted to research on materials and other (solid) substances. The two institutes shared a number of technical facilities and also had a joint large library and canteen.

The boundary conditions for working at the Max Planck Institute and those for working at the University are distinctly different. This is not only obvious considering the financial and personnel means offered to a Max Planck Director; I wrote about that before. It also becomes clear visually. I perceived it time and again when I was at the University for a meeting, mostly the monthly faculty board meeting: not only the outside and inside of the Max Planck Institute are more attractive (cf. Fig. 13.4) and more costly realized, also the quality of the furniture just speaks for itself. Of course, these externalities are not at all directly indicative of the quality of the research that is possible, but they help in generating the atmosphere where high quality can emerge.

Yet, I missed somewhat the more intimate atmosphere of "my" Seestrasse environment, in, almost downtown, Stuttgart. I had come to like my small office in Seestrasse 92 (Fig. 13.2c), and the more informal dealings with the local, supporting technical and administrative services, which of course were distinctly smaller there. We had no canteen in the Seestrasse. So we bought our lunch at the butcher's shop nearby, at the corner of the Seestrasse and the Wiederholdstrasse: more expensive than in the canteen in Büsnau but of better quality. When I arrived by train in Stuttgart, I simply could walk (hillside upwards) to the Institute, and arranging some private business in the city was also easily done by feet (hillside downwards). But, of course, these musings are of no significance as compared to the enormous advantages offered in the new building in Büsnau.

Life and living

Moving to Stuttgart I needed a place to live. As it happened, within the building at Seestrasse 92, a service apartment on ground floor had become available and, after some renovating, was offered for temporary use to me (I paid a reasonable rent). This meant that, for the first and only time in my life, I actually lived in the building where I performed my daily work: my office was immediately above (one floor higher up) the living room of my small apartment. In a way this was comfortable (no traffic jam ever to come in the way upon going to work...), but not preferable for a long time: some spatial distance to professional life is desirable for one person's well-being.

I began to look for a house, next to the daily professional obligations. Not only because of limited time, this is not easy; the more as I insisted on a place to live at relatively close distance to the institute, i.e. in Stuttgart, and houses in Stuttgart are expensive.... Finally I could buy a small house in Botnang. Botnang, once an independent village, now is part of Stuttgart, located at its western side and surrounded by forest. It is considered to be the part of Stuttgart with the best atmosphere, since the fresh air for Stuttgart enters, due to the frequent more or less western winds, Stuttgart via Botnang. This was an argument of importance to me, as I already had become familiar with the bad air and, especially the in summer, sticky atmosphere in downtown Stuttgart.

Already only because of the geography of Stuttgart, with its ascending and descending streets, I realized that, in flagrant contrast with my situation in Delft, taking a bike for going to and fro the Institute would be practically impossible (if one was not prepared to take a shower immediately afterwards). So I also bought a car.

Fig. 13.6 Marion and Eric in Eric's office at the Institute in 2014. On the wall behind Eric two of the five recrystallized aluminium sheets inherited from Burgers can be seen (defocussed and thus blurred) (see Chap. 10 under *My contact with Burgers* and see Fig. 10.1)

My secretary, Marion, impressed me right from my start in Stuttgart. After her years serving my predecessor, Professor Predel, she now fully embarked on the project of developing a new department and supported me in all possible ways. She was extremely able in getting things done, either circumventing possible barriers and also by involving the right persons of the Institute, which was of great importance in my beginning time. I liked her more than as a close co-worker and we found out that such feelings were mutual. It was not easy for us to admit this. In the end, after having overcome some barriers, we married and are very happy to have each other till this very day. She remained my secretary and professionally served me and my department more than skillfully and tactfully through the years. Her loyalty is limitless. I cannot imagine life without her (Fig. 13.6).

The house in Botnang was situated on a slope, close to its upper end. Upon entering the house one found him/herself at its highest floor where the large living room was. One floor lower the sleeping room was situated, overlooking, at the backside of the house, what I have called, "the valley of Botnang", with its completely built slopes. At this backside of the house the long garden stretched downwards along the hillside. Marion and I liked the house and enjoyed living there. Fresh on my mind are memories of sitting behind my writing desk at night, at the backside of our living room, where, when I looked out of the window, could see the lights of Botnang one after the other becoming extinguished as time progressed. Chicco, our cat, would come and jump on my writing desk, lay there stretched out, possibly across the keyboard of my computer, and spur. We spent many hours together in this way. If Marion was not already in bed, she was at close distance at the other side of the living room, thus easily allowing contact of the three of us. The atmosphere in this house can only be described as harmonious and homely.

Yet, we moved away because of a number of reasons. Taking care of the sloping garden at the backside of the house was a burden. We had a gardener who helped us regularly, but this was found to be no solution for the days when we would be too old. We decided to act with foresight and look for an apartment that was comfortable enough, a.o. reachable by elevator, to be our permanent residence (so to speak till our death). The awkward relationship with our neighbour, described above, was an additional pressure. Finally, we noticed that the house had reached a stage where considerable renovation was necessary.

After a long search we at last bought the penthouse of a newly built house for 5–6 parties. The house was located in Sonnenberg at the south-western side of Stuttgart. We suddenly had much more space, a lot of balcony/terrace and we enjoyed an impressive panoramic view over Kaltental and Vaihingen.

Unfortunately, we never felt as much at home as in our previous small house in Botnang. Sonnenberg appeared to be a rather "dead" part of Stuttgart, at least as compared to Botnang, which had life and atmosphere by its own and a small but lively centre, down in the "valley of Botnang". I now had a separate and larger home office, what pleased me, but this room was "at the other end" of our penthouse and had as disadvantage that communication of Marion and me was no longer easily realized: the largest part of most evenings we were a separated couple, which was not of our liking. It seemed that even Chicco had liked the house in Botnang more. In the evening she had to decide between me and Marion, which troubled her markedly. Moreover, she had now become old: if she came to me in my home office and made clear, by meowing at my feet, that she wanted to be on my writing desk, I had to elevate her, as she no longer had the power to jump. After some further years she became serious ill and eventually the veterinarian came to us and we concluded her life peacefully, while she was lying in my arms. For Marion and me this was a caesura of lasting impact.

A long series of smaller and larger disagreements between the parties in the house arose. These were incited by usually two of our co-owners, who wished continuously to bring about costly modifications to the house which in my and Marion's eyes were unnecessary and which primarily served the interests of our co-owners. As we possessed the equivalent of about a quarter of the whole house we also had to pay, in proportion, most of all owners for any modification performed. We experienced great difficulty in serving our interests by ourselves. This only improved after a friend, a judge of profession, helped us, as I remarked above. However, at that stage we were already no longer willing to accept these minor and major quarrelings for the rest of our lives. We decided to leave near to the moment of our professional retirements. Thus in 2016 we moved to a new apartment at the very outer edge of Heidelberg, looking over the fields in the direction of the mountain ridge of "die Haardt", at the opposite side of the Rhine, which we can see at days of clear sight. Here Marion and I still live in peace and to our contentment.

Moving with a complete research department and its equipment from one location to another one is an enormous endeavour: it took about six months in the new building in Büsnau before all important instruments, as the calorimeters, the dilatometers, the X-ray diffractometers and the surface analytical (Auger and X-ray photoelectron spectroscopy (XPS)) equipment, could be operated again more or less to our satisfaction.

The years of research by the department in Büsnau, i.e. since 2002 till my formal retirement in October 2016, were very successful ones. The department had reached its definitive shape and size after the move from the Seestrasse to Büsnau and ran smoothly as a well-oiled machine. The start, more or less from scratch, in the Seestrasse, with associated disturbances, now lay behind us and the research program and structure of the department fully reflected my ideas.

A materials scientist, in order to stay at the forefront of science, cannot avoid to also (further) develop methods and techniques of investigation. This is normally not mentioned explicitly if one presents the results of his/her research on, as in my field of science, material behaviour.

Therefore, as a statement of tribute, it is here recalled that scientists of my department were the first to develop and apply an X-ray diffractometer equipped with a rotating anode source in combination with a single-reflection, two-dimensional, collimating X-ray mirror[10] to produce a *parallel* beam of X-rays of *high intensity* (Udo Welzel and Markus Wohlschlögel). The intensity achieved was that high that it was compatible with the intensity of early synchrotron radiation sources, which allowed the detection of very weak reflections hitherto unnoticed.[11] The parallel nature of the resulting X-ray beam was of special use upon measurement of stress in material systems (as thin films).

A similarly significant instrumental development was due to a second group of scientists of my department (Lars Jeurgens, Zumin Wang and David Flötotto): in the course of years, a custom-designed, unique, ultra-high vacuum (UHV) system was built and enlarged for specimen processing and in-situ analyses. The system consisted of interconnected UHV chambers for thin film growth (by vapour deposition and by oxidation), in-situ angle resolved X-ray photoelectron spectroscopy (AR-XPS), in-situ scanning tunneling microscopy (STM), real-time in-situ ellipsometric spectroscopy (RISE), and real-time in-situ substrate curvature measurements for determination of the stress in the (growing) films using a multi-optical stress sensor (MOSS). This complicated and comprehensive equipment for example allowed acquisition of deep insight into the growth of and phase and stress formation in thin films (e.g. metal-oxide films).

Therefore, at those places where earlier in this booklet or in the following, interesting or even important results of research have been or, respectively, are presented, it must be realized that such outcomes have only been possible by, at the background, also the dedicated methodological and instrumental developments at high level, as illustrated by the examples in the preceding two paragraphs.

---

[10] The general reader of this booklet needs not to fully appreciate the meaning of these technical terms (and those in the following paragraph of the main text). For those interested here: The single-reflection collimating X-ray mirror is a multilayer optic of parabolic shape *with respect to two mutually perpendicular planes* and thereby the collimation is achieved, both "horizontally" and "vertically", using a single (instead of two consecutive) Bragg reflections.

[11] If we would have had this equipment to our availability at the Laboratory of Metallurgy in Delft in the early nineties of past century, likely there would have not been the need to apply for measurement time at and to travel to one of the few storage rings worldwide for the production of synchrotron radiation (in that case to the United Kingdom; see "*Iron Nitrides and Iron Carbides; Tempering of Steels and Pre-Precipitation Processes*" in Chap. 10).

Interface and Surface Energetics;
a Materials Science Approach; wrong "Diagnostics"

Nanosized materials have drawn enormous interest in last decades. Thin film systems, in their simplest form composed of a (very) thin film on top of a thick substrate, provide the classical example of nanomaterials. There are many *technological* reasons for the interest in nanomaterials: for example, they can dominate the performance of microelectronic and photovoltaic devices, coatings, gas-sensors and metal catalysts. The great *scientific* interest for nanomaterials is derived from the following recognition:

Nanomaterials possess a relatively large amount of atoms associated with interfaces and surfaces (i.e. relative to the amount of atoms in the bulk). Therefore the contributions of the interfaces and the surfaces to the total energy of the system cannot be ignored (as is usually done for bulk materials). This can lead to the observation of unexpected, unusual and new phenomena which allow deep insight into the principles governing material behaviour. Against this background a large part of the research of my department has been devoted to nanomaterials.

Atoms at the surface of a material, as a pure metal, have chemically less desirable surroundings than atoms in the bulk of the material: they are less intensively bonded as there are less (bonds with) neighbouring atoms. Therefore surface atoms have a higher energy than bulk atoms: the surface energy is defined by this excess energy per unit area of surface. A similar remark can be made for the interface between the grains (single crystals) that constitute a polycrystalline material, as a pure metal. At the interface the atoms are in a state of less ideal chemical bonding. The interface energy is defined as the excess energy that interface atoms have, as compared to bulk atoms, per unit area of interface. The above explanation immediately suggests that the surface and interface energies are positive. However, this is not generally true. Consider the interface between a pure component of element A and a pure component of element B. If atoms A and atoms B like to bond with each other, it is likely that the atoms "feel" better at the interface of component A and component B, than in their bulk environments. Then the interface energy will be negative.

The unequivocal, experimental determination of the surface energy of solid substances is extremely difficult. The direct, unequivocal, experimental determination of interface energies is impossible; indirect routes are highly problematic. Purely theoretical calculations (as first principal calculations) require detailed knowledge on the surface and interface structure (such as the precise coordinates and types of atoms at or close to the interface), which data are normally unknown. This situation has not changed for at least the last 50 years or so. For a materials scientist this state of affairs is highly unsatisfactory. Therefore semi-empirical approaches, where theoretical concepts are combined with experimental data that are available or retrievable, can become important.

Shortly after I had completed my Ph.D. project and had started in the group of Korevaar at the Laboratory of Metallurgy in Delft (see Chap. 8), at a Dutch meeting of materials scientists at the University of Amsterdam, I attended a lecture by Professor Andries Miedema, who at the time was extraordinary Professor at this university, next to his main job at the Philips Research Laboratories, the NatLab (where he would be the deputy director "Basic Physics and Materials Science" after 1980). This was the first time I heard about Miedema's approach to calculate thermodynamic quantities, for which no experimental data are available, as, in particular, the enthalpy of formation of binary solid solutions. His model, which he developed (further) in cooperation with his co-workers Rob Boom and Frank de Boer, is called the "macroscopic atom model".[12] At that time in Amsterdam I could not know that we would incorporate that approach extensively in our much later work on the assessment of interface and surface energies.

The first paper on the "macroscopic atom model" should have been published in a first class journal as *Acta Metallurgica*, a leading journal in the field of materials science, or the *Physical Review*, where groundbreaking work in the field of materials science is also often published. However, Miedema's work was rejected, although he showed how well his predictions fitted with reality. Miedema's ideas did not fit with the concepts used by the established elite of scientists in this field of science; Miedema could not persuade his referees (also see at the end of Chap. 6). Therefore Miedema first published his work in the Philips journal *Philips Technisch Tijdschrift* (in 1973) and later (in 1975) in the less well known *Journal of Less-Common Metals* (this last journal was later renamed as *Journal of Alloys and Compounds* and has gained a better reputation). In the following years until the present day the "macroscopic atom model" acquired enormous popularity, because of its successful predictive power.[13]

---

[12] The essence of the semi-empirical, "macroscopic atom" model is that an atom is considered to be a piece of material (in the model this piece is a Wigner–Seitz cell) with the macroscopic properties of that material. Strictly considered this is incorrect: only an aggregate of atoms of one kind exhibits macroscopic properties which identify the elemental material concerned. Bringing two such "macroscopic atoms", each of a different element, into contact (as in a solid solution, which is the case originally considered by Miedema for the assessment of the enthalpy of formation of solid solutions, or as at the interface of two solids of different elements, which is the case of relevance to our work on the assessment of interface energies!) energy changes occur in association with a redistribution of electron density governed by the difference in electron density and the difference in chemical potential of both "macroscopic atoms". Thereby available data on macroscopic properties can be used to calculate (predict) enthalpy of formation values.

[13] The book published in 1988 where the "macroscopic atom model" is elaborated with all available experimental verification available at the time (*F.R. de Boer, R. Boom, W.C. M. Mattens, A.R. Miedema and A.K. Niessen, Cohesion in Metals: Transition Metal Alloys, 1988, Elsevier Science, North-Holland, Amsterdam*) has been cited by now about 4000 times (personal communication by Rob Boom, now

Miedema was the leading personality under Dutch materials scientists and physicists. He figured in numerous committees and councils and also was an advisor of the government. He had great influence on the materials and physics research performed at the Philips Research Laboratories (the NatLab; also see at the end of Sect. 12.1) and enjoyed great successes therewith (magnetic materials). At the end of the eighties and the beginning of the nineties he was confronted with the decay of fundamental research at Philips which also cost personal positions (see footnote 18 in Chap. 5). This depressed him. Unexpectedly Miedema died at the age of 58 in 1992.

Miedema's scientific contribution allows to place him into the illustrious sequence *van Laar—Meijering—Miedema*, who all are high caliber Dutch scientists, of (very) different epochs, who contributed essentially to our understanding of the thermodynamics of materials (more specifically alloys) (regarding Meijering, who performed this work also while affiliated at the Philips Research Laboratories (as, later, Miedema), see Chap. 8 under *Professor Meijering*).

Our contribution to this thermodynamic approach was its incorporation in our assessment of surface and interface energies for which no experimental data are available (see above). An overview of these models for the calculation of surface and interface energies is provided by *International Journal of Materials Research, 100 (2009), 1281–1307*. The significance of this work is that we were able to demonstrate that previous interpretations of phenomena, considered to be in conflict with what was expected on the basis of strived for equilibrium, suffered from wrong "diagnostics":

A physician, confronted with unclear illness phenomena, may often be inclined, out of lack of understanding, to conclude that the patient suffers from some sort of "influenza". Similarly, materials scientists, out of non-understanding, often ascribe supposed to be non-equilibrium phenomena to "obstructed kinetics": it then is supposed that the (atomic) mechanism that would lead to the state of equilibrium is "blocked", e.g. because of a temperature that is too low to allow sufficient atomic mobility.

As a student I had followed a lecture series given by Professor Burgers on "Physical Chemistry of the Solid State" (see Chap. 10 under *My contact with Burgers*). He (also) spoke about the oxide film on pure aluminium (Al), which formed naturally at room temperature with a thickness of at most a few nanometers (a nanometer is 10^{-9} m; the size of an atom is a number of 10^{-10} m). This oxide layer was amorphous. Of course we knew that the

---

emeritus Professor at the Delft University of Technology and, by the way, the current possessor of my talar (see above under "*Manners and Mores*")…).

crystalline state of the oxide is the more stable one, i.e. the energy of the crystalline modification is lower than that of the amorphous modification. Therefore Burgers said that the amorphous nature of the observed oxide film was a consequence of slow "kinetics": the low mobility of the atoms at room temperature would obstruct realization of the equilibrium, crystalline modification of the oxide. At that time this was believed by all.

A generation later, using our methods to calculate interface and surface energies, we demonstrated that, for the oxide film on the metal, the sum of the energy of the interface with the metal and of the surface energy is lower in case of the amorphous modification than in case of the crystalline modification. As a consequence the amorphous modification is the *stable* modification for very thin amorphous oxide layers. Beyond a *critical thickness*, which we could calculate with our models, the lower bulk energy of the crystalline modification would become dominant and then the crystalline modification would prevail. This turned around the originally adopted and widely accepted view, as expressed by Burgers in his lecture.

We performed experiments and showed that also our predictions for the critical thickness values, which depend on the crystallographic orientation of the aluminium crystal at the surface of the metal before oxidation, agreed with the determined experimental values for the critical thickness. The critical thickness values as calculated for a number of oxidizing elements as function of the oxidizing temperature are shown in Fig. 13.7. As long as the oxide-layer thickness is smaller than the critical thickness the oxide layer is amorphous and this *is* the stable modification. The relatively large value of the critical thickness for the oxide layer on silicon (Si) is a consequence of a relatively small difference in bulk energy of the crystalline and amorphous modifications for this silicon oxide. Note that not for all metals an amorphous oxide film is predicted to be the stable modification: oxide layers developing on chromium (Cr), copper (Cu) and iron (Fe) are predicted to be crystalline, as indicated by negative values for the critical thickness in Fig. 13.7, i.e. in those cases the differences of the bulk energies of the amorphous and crystalline modifications of the oxide are not small enough.

A similar misconception as expressed in Burgers' time for the amorphous aluminium-oxide layer, was repeatedly presented in the literature for the process of "solid-state amorphisation". Instead of the *crystalline* intermetallic compound expected as reaction product at the interface of specific metals upon annealing, an *amorphous* solid solution appeared at the interface of both metals (for example this holds for nickel/titanium (Ni/Ti), copper/tantalum (Cu/Ta), aluminium/platinum (Al/Pt), magnesium/nickel (Mg/Ni)). This

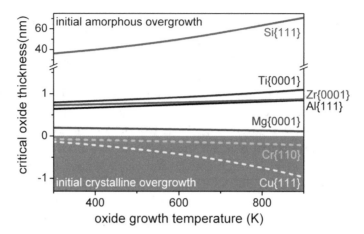

Fig. 13.7 The critical thickness values, for the transition from the amorphous state (below the critical thickness) to the crystalline state (above the critical thickness) for oxide layers on surfaces, of indicated crystallographic orientation,[14] of a number of oxidizing elements as function of the oxidizing temperature (Si = silicon, Ti = titanium, Zr = zirconium, Al = aluminium, Mg = magnesium). Thickness values along the ordinate are expressed in nanometer (nm; 1 nm = 10^{-9} m (m)); the temperature along the abscissa is expressed in Kelvin (K; the temperature expressed in degrees centigrade (°C) simply is the value of the temperature in K + 273). A negative value for the critical thickness, as for chromium (Cr) and copper (Cu), implies that the oxide layers developing on these metals are (predicted to be) always (i.e. for all thicknesses) crystalline (adapted from *F. Reichel, L.P.H. Jeurgens and E.J. Mittemeijer, Acta Materialia, 56 (2008), 659–674*)

phenomenon was ascribed to "kinetic obstacles" for the formation of the crystalline intermetallic compound and/or "anomalously fast" diffusion of one of the components. We could show that the amorphous layer at the interface was energetically stabilized, and thus preferred, due to the interface energy of both crystalline/amorphous interfaces (i.e. at the interface with the metal underneath and at the interface with the metal above) being lower than the interface energy of the corresponding crystalline/crystalline interfaces. This means that up to a critical thickness the amorphous layer has a total energy lower than that of the crystalline layer and therefore the amorphous layer is the energetically stable modification.

A spectacular application of the above concept of the role of surface and interface energies on the stability of amorphous layers was the prediction and experimental verification of the strong temperature dependence of the

---

[14] The crystallographic orientation of the surfaces has been indicated in Fig. 13.7 by the numbers within the braces: the so-called Miller indices (for the materials silicon (Si), aluminium (Al), chromium (Cr) and copper (Cu)) and the so-called Miller-Bravais indices (for the materials of hexagonal crystal symmetry, as titanium (Ti), zirconium (Zr) and magnesium (Mg)).

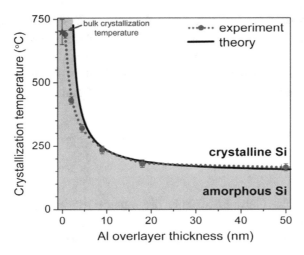

Fig. 13.8 The strong temperature dependence of the crystallization of an amorphous semiconductor as silicon (Si) by thickness variation of an adjacent metal (aluminium (Al)) layer. The bold full line represents the model prediction. The red dots connected with the dashed red line indicate the experimental data. The agreement of the experimental data with the theoretical prediction provides a spectacular demonstration of the role of surface and interface energies in stabilizing nano-sized materials. Note the, as compared to the crystallization temperature of bulk amorphous silicon (about 700 °C), drastic reduction in crystallization temperature for increasing thickness of the aluminium layer: from about 700 °C to about 200 °C for an aluminium layer thickness of about 15 nm (nm); 1 nm $= 10^{-9}$ m (m). Thus the low temperature production of nanoelectronic devices based on crystalline silicon becomes possible (adapted from *Z.M. Wang, J.Y. Wang, L.P.H. Jeurgens and E.J. Mittemeijer, Physical Review Letters, 100 (2008), 125503*)

crystallization of an amorphous semiconductor as silicon (Si) by thickness variation of an adjacent metal layer (Fig. 13.8; *Physical Review Letters, 100 (2008), 125503*). This result has an important technological consequence: The drastic reduction in crystallization temperature for increasing thickness of the adjacent aluminium layer makes possible the low temperature production of nanoelectronic devices based on crystalline silicon: the aluminium layer thickness is a tool to sensitively tailor the crystallization temperature of the silicon.

Against this background it was natural for us to suggest that the unusual amorphous nature of the silicon–nitride (Si$_3$N$_4$) particles(precipitates) in nitrided iron-silicon (Fe–Si) alloys was due to a relatively low value of the amorphous(particle)/crystalline(surrounding matrix) interface energy that overcompensates the, as compared to the crystalline modification, larger bulk energy of the amorphous modification for *small enough* particles. The very first stage of precipitation then can be development of amorphous instead of crystalline particles (see above: *"The discovery of amorphous precipitates in a crystalline solid"*). It seems possible that more intensive and more extensive investigation, than done or possible until now, of such very initial stages of precipitation will reveal the emergence of stable amorphous modifications in those cases where a small enough difference in bulk energy of the amorphous and crystalline modifications occurs.

Language

Although I spoke German in daily life outside the institute, including the communication with my wife, inside the institute the scientific discussions and our reading and writing tasks were all performed in English. Moreover, coming to Germany at mature age I no longer possessed the talent to absorb and command a foreign language, as German, as perfectly as possible for a child. I never lost the accent that marks me as a Dutchman. Once I had a meeting at Bosch laboratories in Schwieberdingen, near Stuttgart. I had to introduce myself to the porter at the gate and inform him who my host at Bosch was. After I had spoken, the man looked at me, grinned and said: "Ha, Rudi Carrell!" The Dutchman Rudi Carrell was a famous showmaster in Germany with the unmistakable accent that I apparently shared with him. I accept but regret the deficiencies in my German. To me it appears that my English is still significantly better; at least drafting a text in English is still easier for me.

After so many years of living in Germany, a very disappointing, even annoying consequence has been laid bare to me, as an unpleasant outcome of living with German and English as the languages of use in daily life: my command of my mother tongue has deteriorated to some extent. During the ill frequent times I speak with my relatives in The Netherlands I have regularly difficulties to find the correct words in Dutch; even my grammar has slight deviations. I had believed that one never loses the capacity to express oneself flawlessly and effortlessly in one's mother language. Evidently, this is not fully correct; a dismaying experience.

During my beginning years in Stuttgart the governing language for communication among the Directors at the Institute and with head office of the Max Planck Society in Munich was German. Only later, after more and more Directors of foreign origin had been appointed by the Max Planck Society, English got a role equivalent to German within the Max Planck Society (for example, the letters from the President to the Directors became bilingual). The Board of Directors meetings at the Institute took place initially also in German and the minutes were also in German. As at the time I was the only non-native Director at the institute (with the exception of my colleague Arzt, who is Austrian, but thus also has German as mother tongue) and I did speak German, it was considered no serious problem for me. And of course, my colleagues were willing to immediately switch to English if I desired so. I almost never asked. Yet, as the reader may imagine, I now and then did miss some, alas sometimes also significant, remarks or phrases, especially if the speed of speaking was high. Although a non-native speaker may be able to express him/herself well in a foreign language, it should be realized by his/her discussion partners who are native speakers that still serious misunderstandings can happen. I have experienced this more than once.

It should be a matter of course that once one has decided to live for significant time (years) in a country where the natives speak a different language, that one then has the also moral obligation to learn that language. There is no other way to become familiar with its country culture and character. Moreover, it facilitates daily life outside the working place/institute in an essential way. It then surprises to observe that the one or the other newly appointed, foreign Director at the Institute effectively refuses to learn German. This will not harm his/her science, but a lack of willingness to integrate into the German society may not be in line with what is best for the Max Planck Society.

Relationship with the University

Occupation of a Chair implies that a full load of teaching obligations rests on the Chair holder. As compared to Delft, the only essential difference for me in Stuttgart was that I had to lecture in a different language. I believe the quality of my German sufficed.

I liked the contact with the students that I experienced as stimulating. Now and then some question led to unexpected, deeper insight into a specific problem by rethinking of it, questioning myself and looking in a different way on the same difficulty. I normally did not use prefabricated slides as in "power point presentations", because lecturers using these often presented too much material in too little time and to a significant part the lecturer just told what was on the slides; then a student really can better go home and study some textbook. In that case I thus agree with those students who say they need not attend lectures.

My objective was never to aim at the largest amount of material presented but to arrive with the students at understanding of the topics dealt with. I did it in the classical way: writing what I thought was essential on the (black) board and lecturing slow enough, sometimes in the form of a discussion with the students, so that the students could make notes by themselves. Just sitting in the audience and listening to an even beautiful presentation is much less effective. The handwriting action involves activation of various parts of the brain that allow better impregnation and grasp of the stuff presented. I was one of the few Professors favouring this teaching method. It would be most optimal if the students would elaborate their notes when back at home behind their writing desk while additionally consulting a textbook, as I had done as a student, but I was realist enough to know that that would be the exception.[15]

Two Directors at the Institute had a similar position as I had: they also occupied a Chair at the Faculty of Chemistry at the University of Stuttgart: Fritz Aldinger and Eduard Arzt. The three of us carried the study course Materials Science at the University. We had a pleasant working relationship. Usually we met on the occasion of the so-called "Kollegialprüfung" (= "collegial examination": examination before a panel of examiners). After having followed basic courses given by me, Aldinger and Arzt, a student after its first year was examined about these courses, i.e. a wide ranging field, in one oral examination by the three of us. The examination was much feared and many failed the first time. Yet, often an enjoyable discussion ran and, perhaps unexpected for an outsider, also the examiners were genuinely animated, which could lead to interruptions by the colleagues in an on-going discussion between the candidate and one of us. The "Kollegialprüfung" was meant as the test of allowance for continuation of the study. If one passed successfully one had achieved something and it felt this way. Perhaps surprising, the students were not opponents of the "institution" "Kollegialprüfung".

After quite some years of the routine described in the preceding paragraph had passed, late in 2007 both Aldinger retired and Arzt accepted a leading role at the Leibniz Institute of New Materials in Saarbrücken, and thus both left the Institute. Because of reorganization at the Institute and a changing relationship with the University, quite a number of years the Chairs of Aldinger and Arzt remained empty or were temporarily occupied by an emergency appointment (of a retired Professor, who helped out). It took until late in 2013 before successors were appointed on the Chairs of Aldinger and Arzt… Hence, because of lack of a possible replacement, my stay as the Dean of the Study Course Materials Science was extended to eventually more than 10 years. This became in particular a burden when we had

---

[15] The reader may try him/herself: having to write up *oneself*, in one's own words, the essence/explanation of a scientific phenomenon, also a derivation of an equation, almost automatically and immediately brings about fathoming of the science involved.

to transform the Study Course into a Bachelor/Master programme on Materials Science, as a consequence of the Bologna agreement.[16] In the absence of my two colleagues this task had to be executed largely by myself and my co-worker Ralf Schacherl. It remains unclear if the Faculty of Chemistry ever realized that this job had to be and was done by only two persons.

Aldinger, Arzt and I had the same opinion about what materials science is (see Chap. 8 under *Materials Science*) and we defended that position within the Faculty of Chemistry, where a tendency prevailed to interpret materials science as a part of chemistry, i.e. ignoring the equally important roles of physics and (mechanical) engineering. After the Presidency of the Max Planck Society had made clear that the positions of Aldinger and Arzt (and after my retirement also my position) would not be continued as a conjunction of Director at the Max Planck Institute and Chair Holder at the University (see next paragraphs), the University could decide independently about appointing successors at the Aldinger and Arzt Chairs (and this took many years; see previous paragraph). Now I was the only "materials science professor" in the Faculty of Chemistry and thus the weight of "materials science" in the Faculty had reduced. Thereby my influence on the new appointments on both Chairs was naturally limited. Therefore it cannot surprise that one of the Chairs now was distinctly oriented towards chemistry, less to materials science.

I have always believed in the advantage of joining the position as Director at the Max Planck Institute with that of Chair Holder at the University; I considered this a very favourable and profitable construction that was one of my reasons to decide for Stuttgart (see *Start in Stuttgart* in this chapter). I did not know from the start that within the Max Planck Society also second thoughts existed about these double job positions, which for the rest were

---

[16] The Bologna agreement (1999) strives for unification of standards and levels (degrees) of university educations in Europe. This would lead to higher mobility of students so that an education began at university A in country C can be continued without barriers and loss of time, i.e. without more ado, at university B in country D. The University of Stuttgart (as held for all universities in Germany) was very late in adopting this system. We introduced the Bachelor/Master programme on Materials Science in 2008. The language of use in the Bachelor programme remained German; the language of use in the Master programme became English. Quite a number of years earlier the adaptation to the Bologna agreement had already been established at the universities in quite some countries of Europe, including The Netherlands and there the language of communication in both the Bachelor and Master courses is English.

One may not favour using English also in a Bachelor course. Native students may have difficulties if the lecturer speaks English, since this adds to the barrier the students experience in any case upon entering the university with its completely different way of "teaching and learning" as compared to the preceding "school". Moreover the "English" of the lecturers may be of limited quality. Therefore in Stuttgart we maintained German as the language of communication in the Bachelor programme. As preparation for use of English both as the lingua franca of science and in the Master programme, we did advise English language textbooks for studying in also the Bachelor programme.

not restricted to Stuttgart alone. This potential ambiguity may be understood upon considering the Humboldt principle and the Harnack principle.

The Harnack principle is named after Adolf von Harnack, the first President of the Kaiser-Wilhelm-Gesellschaft (refounded as Max-Planck-Gesellschaft in 1948). Outstanding scientists ("*the brightest minds*") are appointed as "Wissenschaftliche Mitglieder" (= "Scientific Members") whom are given the means to build up a department of which they become its Director. The Director alone decides on the research objectives, the research methods and the strategy followed. An interdisciplinary and independent character as well as the selection of research areas, that are both new and promising, are distinctive parameters determining the choices of the Max Planck Society.

The Humboldt principle is named after Wilhelm von Humboldt, a Prussian statesman, who, in the nineteenth century, formulated four principles of (academic) education. The essence of his ideas is that (academic) education is realized by science as vehicle: the unity of research and teaching should be the leading principle for (university) studies. This idea can still be considered as a guiding principle for universities.

The confrontation of students with a scientist active at the frontiers of science is most stimulating and fruitful for them, by the transfer of attitude and way of thinking of an active, independent, high caliber scientist. I have experienced this myself, as student and as becoming scientist (see my recollections of Burgers (Chap. 10), Meijering (Chap. 8) and also Bennema (Chap. 5)). Moreover, I am well familiar with the reverse effect too: teaching science reveals the own deficiencies in understanding, for example as exposed by questions of students, e.g. out of their "innocence". This can lead to new and deeper insights and new ideas. Therefore it is my conviction that especially Max Planck Directors, representing a scientific elite, should be involved in university teaching too. This does not represent a conflict with the Harnack principle, as long as the direct involvement in research is not endangered. On the contrary, as I have argued above, the science of the Director profits from his activity as university teacher.

The connection of Max Planck Institutes to universities has had its up and downs in the history of the Max Planck Society. At times the involvement of Max Planck Directors could even be rather intense. My pre-predecessor as Director at the Institute, the longtime Managing Director of the Institute, Werner Köster (see above "*The Max Planck Institute for Metals Research (MPI-MF)*"), even acted as Rector of the University and my immediate predecessor, Bruno Predel (see above), was for some years Dean of the Faculty. Also knowing this, the above text may make clear how much I regretted that the conjunction of Director at the Institute and Chair Holder at the University was dissolved, as the acting Presidency considered the cooperation on that basis as too intense.

The ambiguity of the discussed relationship Max Planck Institute—University is underlined by the need of MPIs of at least some cooperation with (local) universities in order that Ph.D. candidates, who did their research at a MPI, can be lent the desired doctor's degree. It may be assumed that most Max Planck Directors will be pleased with an even only extraordinary Professor title, so that they can act as promotor themselves, which requires cooperation with a university. Certainly, there is a lot to gain for Germany and science(!) by a good relationship of the universities and the Max Planck Society.

In my Farewell Lecture in 2016 I returned to this matter. I cited from the minutes of a long ago meeting at 14-7-1931 of the Board of Directors and the Board of Trustees of the Kaiser Wilhelm Institute for Physics in Berlin (note the names of the participants in the discussion):

"Aus den verschiedensten Gründen sei es wünschenswert, das Institut in eine enge Beziehung zu der Berliner Universität zu bringen." .... *"In der Debatte, an der sich die Herren von Laue, Nernst, Richter, Haber, Franck, Niessen, Glum, Einstein, Konen beteiligten, wurde der Plan der Kooperation mit der Universität lebhaft begrüsst...."*

(= *"For various reasons it is desirable that a close relation is realized of the Institute with the University of Berlin"* ......*"In the discussion, with participation of the gentlemen von Laue, Nernst, Richter, Haber, Franck, Niessen, Glum, Einstein, Konen, the plan of cooperation with the University was greatly welcomed...."*).

Conferences

Communication among scientists plays a substantial role in the progress of science. Nowadays we are no longer dependent on intrinsically slow surface mail, as for example applies to Lorentz' days, where letters were of vital importance (see Chap. 3 under *"On Lorentz and Points of Contact; the Lorentz Factor"*). Conferences, i.e. gatherings of groups of scientists where scientific results are presented, can be of crucial significance as they establish the personal contact, apart from being informed about new and/or unknown science in the field covered by the conference.

As a younger scientist I visited regularly a number of huge conferences, where the number of participants easily reached many thousands and where numerous, say more than 10, parallel sessions were held. This was a perfect means to get an overview of the entire breadth of my field of science and see who was important. Further I got experience in presenting a scientific lecture about my work at an international meeting of significance. Thus I learned a lot in a short time and made acquaintance with relevant colleagues. However, through the years these positive effects become reduced upon repeated attendances at these big, yearly or two/three-yearly held conferences. Moreover, really new information on such huge meetings is nowadays presented only rarely, also as a consequence of the currently high speed of electronic publication (possibly also in unrefereed state in specific www archives providing "open access"). Therefore, in my later years I did not feel a strong need to attend such colossal meetings, but did encourage my young Ph.D. students and post-docs to attend and participate in such conferences, supposing they would make the same useful and positive experiences as I once lived through.

The following anecdote may serve as an illustration of how important personal contact initiated at a meeting can be for the scientific process. Once, in the beginning of the nineties, I returned from a conference in Germany. At the conference I had met Jan Slycke, a Swedish scientist, somewhat older than I was. We knew each other since an earlier meeting in Birmingham in 1981, where Jan had presented interesting work on void formation in steel upon carbonitriding iron-based materials, which was a part of his Ph.D. project work that he had completed not long before. At the same meeting I had presented our "nitrogen pump", observed as a consequence of void formation in a foil of pure iron upon nitriding in the austenite-phase region. Thus we had a lot to discuss already then. Since then it has been clear to me that Jan is a much gifted scientist, who moreover is modest and sympathetic. Jan now worked at the SKF Engineering and Research Centre in Nieuwegein, The Netherlands. As both of us, after the meeting in Germany, had to return to The Netherlands and had decided to travel by train, we now sat together in a compartment of the train heading home. The trip would take about six hours, so we had time to talk.... At the meeting we had just attended it had become clear to me that misunderstandings about the meaning of (thermodynamic) quantities as activities and chemical potentials of, especially, gas components dissolved in solids, as metals, were rather common. Jan had the same impression. After some extensive exchange between the two of us, how precisely activities and potentials in these specific cases had to be defined (several possibilities are possible, adding to the confusion) and interpreted, and noticing that also in our discussion misunderstandings and confusion arose and were resolved, I proposed to write a review paper together about this topic. Still in the train we started to draft, step by step, how we had to devise the treatment in that paper. Back in Delft I immediately prepared a first draft and sent it to Jan. Next we added extensive examples demonstrating the consequences for the control of various metal-gas reactions. The resulting review paper, so not offering genuinely new science but presenting an explicit treatment we had not seen before in the corresponding literature, has served many very well. Many years later (I was approaching retirement and Jan had just retired), a return to our cooperation took place: Jan, I and Marcel Somers embarked on a project to present a comprehensive overview on the fundamentals of the thermodynamics and kinetics of gas and gas–solid reactions. It resulted in a chapter in a book that required more than 100 pages of fine print. Completing this work had almost driven me to "madness", as I had many other obligations as well and usually had to concentrate on this specific job in the (middle of the) night, but the intensity and fruitfulness of the almost daily contact with Jan by email during some of such weeks has been a most rewarding time in my scientific life.

It appears that conferences decades ago had on average a much larger impact on the progress of science than nowadays; the quality and significance of the presented contributions were pronouncedly higher. The example that comes to my mind is, of course, those Solvay conferences for physics which were once chaired by Lorentz, now more than a century ago; conferences not at all of massive nature, indeed. The general decay of the quality of the conferences in the last fifty years or so of course has the same origins as described in Chap. 6 for the decrease of quality of the average paper published during the same period and needs not be recounted here.

Conferences dedicated to a single, not wide ranging theme can be highly valuable as they allow a more in-depth confrontation with existing problems and allow intense discussion among the experts of the world. I have been the organizer of a number of such conferences, as on size-strain analysis, residual stresses and metal-gas reactions. From a scientific point of view these meetings can be very useful and can establish a worldwide, scientific community: to start with, they are not yearly held, but, for example once in three–five years, implying that really significantly new stuff can in principle be presented, and the number of participants typically is 150–200. The value of such meetings can be further enhanced if (a selected number of) the scientific contributions after the meeting is reworked as a genuine manuscript, refereed and collected in a book that then is the scientific result of the meeting. This rather rare procedure was applied to a conference I organized together with my colleague Paolo Scardi in Trento, Italy, in December 2001. The (hand)book resulting from this meeting, by a process as described immediately above, was *"Diffraction Analysis of the Microstructure of Materials" (Springer, 2004)* and is still cited.

Conferences allow more than scientific exchanges. The local environment of the conference site, being situated at a usually appealing place in a foreign country, for most of the conference participants can exert an additional attractive force for attending the meeting. Then there is the "conference reception", the "conference excursion" and the "conference dinner" and, often, the "accompanying person ("ladies") programme". These happenings all allow the development of social relations, which is a positive result of a conference (see above). However, one cannot escape the impression that these "attractions" can also provide the overriding motivation to attend a meeting and, unfortunately, in the many announcements for conferences and meetings that I receive, the exotic location selected and the many "sideshows" offered in some of the (electronic) flyers make clear that the science of such a meeting is of lesser importance. And what to think of conferences organized by non-scientific organizations just aiming for financial profit?

I end this consideration about the (ab)use of conferences with a personal recollection indicating how a small, non-scientific experience, in association with conference attendance, had a lasting outcome in the life of Marion and me. In 2010 I attended an international conference on residual stresses in Riva del Garda, Italy. I had to present a plenary lecture at the meeting and not much time was left for private exploration of Riva del Garda, beautifully situated at the northern shore of Lake Garda, and its surroundings. However, Marion accompanied me on one of my walks from the hotel to the conference. Thus we passed along a gallery presenting an exhibition with paintings by Albino Bombardelli. Looking from the outside through a window we were immediately interested in these paintings. We entered and spent some time there. Bombardelli is not only a painter but also a musician (piano player), which is reflected in his paintings as well. The Italian atmosphere and the style of his paintings were very attractive for us (see for yourself at: https://www.youtube.com/watch?v=o4SJdl3HDs8). Especially a rather large painting of a man with a black cat on his right shoulder, the leading painting of the exhibition, exerted a strong appeal on us, also because it reminded us of our beloved cat, Chicco. We took a flyer of the exhibition with us. We played with the thought of buying the painting. However, we shied away from asking the price and, in any case, we couldn't take the painting easily with us back home, because we travelled by train. Years passed. In 2016 we moved to our new home in Heidelberg. I came across the flyer of the Bombardelli exhibition in Riva del Garda. We now had the "courage" to establish contact with the painter himself via email, which we eventually succeeded in after some detours. He informed us (via an interpreter; Bombardelli does not speak or write English or German) that the painting of the man with the cat was still in his possession and that he was willing to sell it to us. As response to an assertion by me he confirmed that the man on the painting was he himself in younger years. We agreed a price and after I had transferred the money to him he sent the painting to us. The painting now hangs on a wall in our living room above the grand piano. Until this very day it has not ceased to please and enjoy us every time we look at it and, of course, that day in Riva del Garda is a very vivid memory.

Internal Stresses and Thin Films; Surprises of Nature

Solids can be subjected to *internal* stresses, also if there is no external load acting on them. This statement can be made understandable by briefly considering a simple example that serves as introduction to what is to be told here.

Consider a thin solid film on top of a solid substrate. Upon cooling both the film and the substrate will be subjected to thermal shrinkage. The film, in the absence of the substrate underneath, decreases its lateral size less or more than the substrate in the absence of the film on top: a volume misfit of film and matrix then exists. Because the film and the substrate are forced to stay "glued" together, upon cooling then a state of stress develops in the system of film on top of substrate. For the case that the film, if unconstrained, experiences a lateral shrinkage less than the substrate, it is easy to see that in the film a stress parallel to the surface develops that tends to compress the film parallel to the surface, i.e. a *compressive* stress (Fig. 13.9).

The film/substrate system considered above remains in mechanical equilibrium. This implies that the developing state of internal stresses is self-equilibrating. Therefore, the stress in the film parallel to the surface, that tends to *compress* the film parallel to the surface, is compensated by a stress in the substrate in the opposite direction, also parallel to the surface, that tends to *extend* the substrate parallel to the surface, i.e. a *tensile* stress (see Fig. 13.9).

The stress in thin films needs not be constant as function of depth in the film. It can greatly vary over a very small depth range. An extreme case is shown in Fig. 13.10: In a very thin tungsten (W) film of only 50 nm ($= 50 \times 10^{-9}$ m) thickness, corresponding with a row of about 185 tungsten atoms, an extraordinarily steep stress-depth profile prevails: the (compressive, as indicated by convention with a negative sign) stress changes from −2800 MPa (= −2.8 GPa; 1 GPa is about 10000 atm)) near the surface to about −1000 MPa (= −1GPa) at a depth of only 25 nm, i.e. a huge stress change of almost 2 GPa (= 20000 atm) occurs over a depth range covered by only about ninety tungsten atoms in a row. This is a colossal stress gradient (= stress change per unit of depth) indeed.

There is a sad note tightly connected with the unique result shown in Fig. 13.10. The first author of the paper, from which this result has been taken, is Yener Kuru. Yener came as a Ph.D. student from Turkey to my department in Stuttgart and completed in 2008 his Ph.D. thesis on the effects of microstructure and stress on diffusion in thin films. Thereafter he went for a post-doctoral project to the USA. Yener was a man of small posture, very eager and ambitious when it came to science, and had a pleasant, cooperative personality. We kept contact after he had left us. In the USA a brain

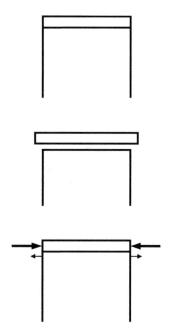

Fig. 13.9 Stress development in a thin layer on top of a substrate upon cooling. In the case shown, the layer desires to shrink less than the substrate. As a consequence a compressive stress parallel to the surface develops in the thin layer (see the arrows indicated next to the film; cf. Fig. 8.3), because the film and the substrate are forced to stay "glued" together. Mechanical equilibrium of the layer/substrate system then requires the simultaneous emergence in the substrate of a (small) tensile stress parallel to the surface

tumor was diagnosed and operated. Yener returned to Turkey, where he got a position at the Middle East Technical University in Ankara. On the basis of his initiative we started working on the project on which Fig. 13.10 is based; he in Turkey, Udo Welzel and I in Stuttgart. During writing up the results in 2014, a new operation in Turkey was necessary: the tumor had returned. Immediately after his return from hospital he resumed working with us. I have to confess that at that time I didn't well realize neither under which constraints he performed this job nor how hopeless his state of health was; he made no such hints in his emails to me. The paper was published in December 2014 in *Applied Physics Letters*; he expressed great joy about having come that far. Yener lived another year with health problems; he died early in 2016 at the age of 36, leaving his wife and child.

One reason for writing the above paragraph can be described as follows. I have been the doctor father of about 75 young scientists. I see them all before my "eyes". But, of course, I "see" them as they were during their time with me. How do they look now and what do they do? Only about a minor fraction of them I know more or less such facts. But concerning the others? Are they still alive? What happened to Yener makes the last question meaningful and

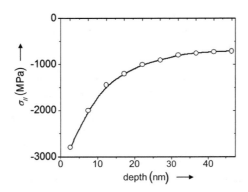

Fig. 13.10 Enormous stress variation over very short distance (in science one then speaks of a colossal stress *gradient*) in an ultrathin (50 nm (nm); 1 nm = 10^{-9} meter (m)) film of tungsten (W): from −2800 MPa (megapascal) (= −2.8 GPa (gigapascal); 1 GPa is about 10000 atm (atmosphere)) near the surface to about −1000 MPa (= −1 GPa) at a depth of only 25 nm, a depth range covered by only about ninety tungsten atoms in a row. By convention compressive stresses are taken as negative, i.e. they are designated by a negative sign before the stress value; tensile stresses are taken as positive (adapted from *Y. Kuru, U. Welzel and E.J. Mittemeijer, Applied Physics Letters, 105 (2014), 221902)*

I have become more aware of it as I have reached an age where dying around me no longer is a rarity.[17]

Nanosized systems, as ultrathin films (of thicknesses, say, a number of nanometers; one nanometer (nm) $= 10^{-9}$ meter (m)) can give rise to hitherto unknown effects. In 2011 David Floetotto, in the course of his Ph.D. project, devoted to stress development in thin metal films during their growth, made a genuine discovery: by performing real-time in situ stress measurements of the stress developing during growth in an ultrathin aluminium (Al) film, it was found that the stress, parallel to the surface, had an oscillating character, i.e. it went through maxima and minima with an amplitude decreasing with increasing thickness. The period of the stress oscillation corresponds with a thickness increase of about three layers of aluminium atoms i.e. about 0.72 nm $= 0.72 \times 10^{-9}$ m (see Fig. 13.11). This observation of stress oscillation was a great surprise to us; something like that had not been seen before. The interpretation of this intriguing phenomenon runs, in simple words, as follows:

The effect is a manifestation of a consequence of quantum mechanics. The free electron energy[18] in a free-standing (ultrathin, aluminium) film has minimal values at specific thicknesses of the film (we speak of "quantum confinement" of the (free) electrons). The film, also *during* growth, strives for realization of these thicknesses, one after the other while growing. The

---

[17] In only one case I acted as the doctor father of a man distinctly older than I was: Aat Voskamp, who was 11 years my senior. Aat worked as an engineer at SKF Engineering and Research Centre in Nieuwegein, The Netherlands. He had an engineer degree from a polytechnic school (this degree was indicated with "Ing." in The Netherlands ("Ing. (grad.)" in Germany), as a distinction with respect to an academic engineer graduated at a university of technology, leading to a degree indicated with "Ir." ("Dipl.-Ing." in Germany)). There was no doubt that Aat (yet) was of academic quality: he had published in the international scientific literature and in the course of time acquired an international reputation in his specialty: microstructural changes during rolling contact fatigue (as in ball bearings, which indicates the relevance for SKF). I knew Aat since my Ph.D. time, as he regularly came to Delft to perform texture measurements. Some 20 years later, in the beginning of the nineties, Aat approached me with the wish to start a Ph.D. project. From this cooperation with me a number of joint papers resulted and then the thesis was successfully defended in 1997, shortly before I left for Stuttgart. This academic promotion was remarkable, also because a special permission of the Delft University of Technology was required, since Aat had no academic qualification up till that time (I had to write an extensive application). This really was a special event, celebrated with a big party, also because Aat in 1997 was already 57 years. His thesis can indeed be considered as the crowning achievement of a lifelong career: a "life's work" (see Chap. 6). With reference to the main text above, the second reason to include this footnote here is that also Aat no longer is amongst the living: he died from cancer in 2014 at the age of 75.

[18] This is the energy of the electrons in the metal which are not strongly bound to an atom but can move "freely" through the piece of metal. These so-called valence electrons move within a "potential well" confined by the surfaces of the crystal (here the interface with the substrate and the surface of the single-crystalline film. The (periodic) arrangement of the positively charged metal ions is ignored: the positive charge is taken to be smoothed out such that a uniform potential results (for some more details, see *E.J. Mittemeijer, "Fundamentals of Materials Science", Second Edition, Springer, 2021, pp. 68–79*).

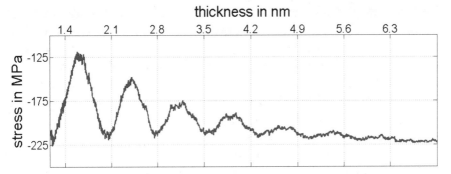

Fig. 13.11 During growth of an ultrathin aluminium (Al) film, the stress, parallel to the surface, shows a surprising, oscillating character, i.e. a series of (compressive) stress maxima and minima is observed in real time, with an amplitude decreasing with increasing thickness (note: by convention compressive stresses are denoted by negative values). (100 MPa (megapascal) is about 1000 atm (atmosphere)). The period of the stress oscillation corresponds with a thickness increase of about three layers of aluminium atoms, i.e. about 0.72 nm (nm) = 0.72 × 10^{-9} meter (m). This phenomenon is a consequence of quantum mechanics: quantum confinement of the free electrons in the ultrathin film (see text) (adapted from *D. Flötotto, Z. Wang, L.P.H. Jeurgens and E.J. Mittemeijer, Physical Review Letters, 109 (2012), 045501*)

establishment of these desired thicknesses is associated with the build-up of elastic deformation energy. However, the film is not free-standing: the film is strongly attached to a rigid substrate. Therefore the film cannot attain its strived for thickness values as the associated changes in lateral dimensions[19] are counteracted by the substrate underneath. As a result a stress parallel to the surface occurs: the stress is *tensile*, when the desired lateral *contraction* of the film is counteracted by the substrate (and thus the desired thickness *increase* (to minimize the free electron energy) cannot be realized; the stress is *compressive*, when the desired lateral *extension* of the film is counteracted by the substrate and thus the desired thickness *decrease* (to minimize the free electron energy) cannot be realized. Thus, this alternation of (desired) expansion and contraction during growth of the film leads to an oscillating nature of the stress during film growth, as observed (Fig. 13.11).

Quantitative results are obtained via the route indicated in the above paragraph. The results show a periodicity for the oscillation of the stress parallel to the surface of the film in agreement with the experimental observation. Also as experimentally observed, the amplitude of the stress oscillation

---

[19] If a bar is pulled along its axis, the lateral dimensions decrease somewhat. If the bar is compressed along its axis, the lateral dimensions increase somewhat. Similarly, if a thin film strives for thickness increase, then its lateral dimensions (i.e. the dimensions in the plane of the film) strive to decrease. If the film strives for thickness decrease, then its lateral dimensions strive to increase.

decreases with film thickness, to become negligible beyond a film thickness of about 6 nm (= 60 Ångstrom = 60 × 10^{-10} m), which corresponds to about a distance covered by a row of 25 Al atoms (25 monolayers). We were excited about this first observation and its explanation and published it soon thereafter in a top journal (*Physical Review Letters, 109 (2012), 045501*).

This was one of those singular experiences revealing secrets of nature that impressed me strongly. These surprises of nature do illuminate nature's overwhelming richness, albeit constrained by the physical laws (Chap. 4).

In the end every phenomenon, also the at present undiscovered ones, will be given a rational explanation, as happened in the case dealt with here.

Books

During the later years in Stuttgart considerable time was spent on a number of book projects which partly run simultaneously. Together with co-workers I acted as Editor of some multi-authored books on topics of our expertise: "*Modern Diffraction Methods*" (Eds.: *Eric J. Mittemeijer and Udo Welzel*), *Wiley–VCH, 2013*; "*Metal-Induced Crystallization*" (Eds. *Zumin Wang, Lars P.H. Jeurgens and Eric J. Mittemeijer*), *Pan Stanford Publishing, 2015*; and "*Thermochemical Surface Engineering of Steels*" (Eds. *Eric J. Mittemeijer and Marcel A.J. Somers*), *Elsevier-Woodhead Publishing, 2015*. The editing work involved selection of potential authors, (continuously) motivating them in order that they fulfilled what they had promised and thereafter reviewing the manuscripts, etc. These projects were time consuming, also because we ourselves drafted distinct parts of these books, in the form of chapters, but these assignments did still not become such demanding of my energy as the book I had authored alone earlier:

Already since about 2004 I worked on the manuscript of a book which would get the title "*Fundamentals of Materials Science*". Actually the idea for this book had emerged already during my time in Delft. On the one hand a number of books about the basics of materials science already existed, but these were usually such simple that they were really only of use for students at the very beginning of their study. On the other hand monographs devoted to specific topics in materials science, e.g. diffusion, mechanical properties, etc., were available, but these presented information on a level more appropriate for students in a (very) end phase of study (and for mature scientists....). Hence, I assigned myself the task to draft a text that could be used in a beginning stage of study, but at the same time could be used at a much later stage of study (say, the Master Course after the Bachelor Course) and in general could provide a basis for successful study of a monograph on a specific topic of materials science.

When I at last had started in Stuttgart with this project, I found out that progress was very slow: it was very difficult to find the time, next to my daily duties, which by themselves already consumed a lot of my evenings/"private time". Burgers' words to me from long ago came to my mind: writing his book on recrystallization "*had almost "killed" him*" (see Chap. 10 under "*My contact with Burgers*").... Finally, after years, and under mounting pressure of the publisher (Springer), the book was completed and appeared in 2010. Soon thereafter a translation in Chinese appeared as well.

The book appeared to be successful. After a few years the publisher asked me to prepare a second edition. In principle I liked the idea, as I wanted to perform some changes and insert significant additions to the book. However, having in mind the burdening time while drafting the first edition, I refused to perform this task while still fully active at the Institute and the University. I started with this undertaking after my retirement. The publisher understood and thus the revised and extended second version appeared in finally 2021. I felt relieved from a burden on my mind that had given me a sense of guilt for a long time.

Editor Experiences

I had been active for a number of years as a member of the Editorial Board of the *International Journal of Materials Research (IJMR)*, when in 2012 I was asked to become its Editor-in-Chief (see what I wrote about the journal in Chap. 6 under "*Editor of a materials science journal*").

Already for a long time the Editor-in-Chief of IJMR has been a Director at the Max Planck Institute for Metals Research in Stuttgart. I was Director at this Institute since 1997/1998. I felt obliged and pleased to continue this tradition after Manfred Rühle decided to retire from this assignment. But I also feared somewhat the amount of associated additional work that moreover could never be postponed (manuscripts keep coming in....).

The Editor-in-Chief of a scientific journal has no small task and carries pronounced responsibility. He/she has the first view on submitted manuscripts and must decide if they fit to the scope of the journal and if they apparently contain sufficient quality in order to let them be refereed. If both criteria are met, the referees (preferably two) must be selected and contacted. This can be a very laborious stage in the handling of the manuscript, because very often potential reviewers refuse to referee because of a variety of reasons.

If the (finally found) referees in their reports express the same opinion it can be easy for the Editor to arrive at a verdict on the manuscript: accept, reject or revise. If the referees disagree in their advice, the Editor must spend considerable time on the manuscript and the referee reports and if he/she then cannot arrive at a decision, an adjudicator must be sought for, who, apart from the manuscript, has access to the referee reports and must advise the Editor on that basis. Only rarely a manuscript is accepted as submitted. Usually revisions are requested (by the referees and/or the Editor) and upon resubmittance of the manuscript after revision a renewed refereeing by the same referees may be necessary. This last step may require even more cycles. Along this entire route the Editor is assisted by software (in my case called the "Editorial Manager"). This is helpful, as all communication is done electronically, of course, but this yet does not reduce the Editor's task very significantly.

In recent years an enormous increase of the number of submitted manuscripts from countries as China, Iran and India, but also Pakistan, Egypt, Indonesia, etc. is evident (see Chap. 6). The English of these manuscripts often is of piteous level, frequently even obstructing getting an idea about what is meant by the author(s). This required an enormous effort by the technical editors of the journal to smooth the language before publication, also leading to a long period of time between acceptance and publication, which is no advertisement for the journal. I became so frustrated of this annoyance that I changed my policy: if a manuscript of bad language arrived, it was immediately to be returned to the author(s) with the request that a professional language/text editing service should be involved to check and polish the language of the manuscript, which was to be proven with a certificate upon resubmittance. This helped significantly.

Before my time as Editor-in-Chief the journal had regularly published issues devoted to the retirement or the 65th, 70th, 75th, 80th etc. birthday of Professors in, usually, Germany or Austria. This meant customarily that a (former) co-worker or colleague of the concerned Professor contacted former co-workers/colleagues who were asked to write an invited paper to be dedicated to the Professor to be celebrated. It was clear to me that these papers were mostly written for the occasion and did not present significant novelty. Such an issue of the journal with these occasional papers did no good to the journal: the papers concerned were scarcely cited. Hence, I decided to abolish this custom.

The task as Editor-in-Chief of a journal of some importance, and of which some elements have been described above, is such comprehensive that in my case it required at least one hour per day behind the computer, which during my active days mostly occurred in the evenings at home. After my retirement as Director at the Institute and Professor at the University, I continued as Editor-in-Chief of IJMR, and with editorial work for a few other journals, until I had become 70. At that age I considered I should retire from also these positions, as one should not continue for many years carrying editor responsibility after oneself has stopped doing research at front level. Accordingly, in February 2020 I stepped down.

"Offspring"

Upon reading this booklet one might forget that not only the direct research results present an image of the activity of a research department. Thus it must be remarked that I have been the doctor father of 56 Ph.D. students in my Stuttgart years. Two of them have become Chair Holders: Eric Jägle (Germany) and Sylvana Sommadossi (Argentina). A few others are active as associate professor at universities in countries as Spain (Gabriel Lopez) and, especially, India (Santosh Hosmani, Sairamudu Meka, Emila Panda).

Quite a number of my former co-workers and post-docs now occupy high academic positions: The following scientists from China (formerly as postdoc or senior scientist active in my department) now are Chair Holders (i.e. "full" Professors) in China: Liu Feng (Northwestern Polytechnical University), Yongchang Liu (Tianjin University), Jiang Wang (Shantou University), Zumin Wang (Tianjin University), Yonghao Zhao (Nanjing University of Science and Technology); Jiang Wang and Zumin Wang spent, consecutively, much more than a few years in my department as senior scientists. Andreas Leineweber, a former longtime senior scientist in my department, now is Chair Holder at the Bergakadamie Freiberg.

I will end this account of my Stuttgart years with a detour to a research project fittingly illustrating the essence of materials science, a project bringing forward fundamental understanding of material behaviour and thereby at the same time allowing significant technological progress:

Whiskering

In the year 2006 I was approached by Werner Hügel from Robert Bosch Ltd. He urged me to start an investigation into the "whiskering" phenomenon. In industry one faced high costs as consequence of "whiskering" exhibited by tin (Sn) films on substrates, as was the case in microelectronic equipment.

Since the lecture course given by Professor Burgers on Physical Chemistry of the Solid State in 1970/1971, that I had followed as a student (see Chap. 10 under *My contact with Burgers*), I knew that needlelike metal filaments can grow out "spontaneously" from the surface of thin metal films, (already) at room temperature. These metallic "hairs" are usually called "whiskers". Whiskering is observed especially for metals of low melting point, as cadmium (Cd), zinc (Zn), and tin (Sn). Whiskers are single crystalline, grow at the bottom by the addition of material from the film, typically have a diameter of 1–10 μm (1 μm $= 10^{-6}$ m) and can reach lengths of several 100 μm (100 μm $= 1/10$ mm); see Fig. 5.3 in "*The Virtue of an Old Memory; the Hartman-Perdok Theory*" in Chap. 5.

In his lecture Burgers explained the occurrence of this phenomenon as the consequence of the development of a compressive stress in the film parallel to its surface (cf. Fig. 13.9): the compressive stress would "squeeze" material from the film out of the surface of the film; a picture that resembles somewhat the experience of observing tooth-paste protruding from the top of a toothpaste tube upon compressing the tube... At the time this "model" appeared very reasonable and as a matter of fact many current publications about the whiskering phenomenon do still adopt the idea of a necessary compressive stress in the film in order that whiskering can take place. But see what follows.

Sn whiskers have very large, unfortunately very harmful, technological consequences. Tin (Sn) films on copper (Cu) substrates are important parts of microelectronic devices as microchips. The Sn whiskers growing from the surface of the Sn(-based) films can bridge the gaps between neighbouring conductors and thereby induce short circuit failure of the electronic component. This leads to enormous costs. The accordingly large technological interest for arriving at understanding for whisker formation and thereby methods for avoiding its formation, explains why I was approached by "Robert Bosch".

We started a project in our department that run until about 2015. The main actors were two successive Ph.D. students, Matthias Sobiech and Jendrik Stein. In the end we arrived at a comprehensive understanding of the whisker-formation process, which can be summarized in simple words as follows (also see "*The Virtue of an Old Memory; the Hartman-Perdok Theory*" in Chap. 5).

In thin films the grain(=crystal) boundaries are mostly oriented largely perpendicular to the surface of the film, leading to columnar grains(=crystals). The internal stress arising in the tin film on a copper substrate is due to the development of an intermetallic compound (IMC), here Cu_6Sn_5, at the original tin/copper interface, within the film and preferably along the grain

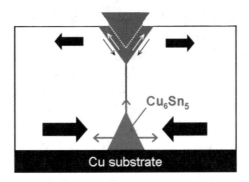

Fig. 13.12 Model for whisker formation: tin (Sn) whiskers grow from the surface of a tin film on a copper (Cu) substrate (see Fig. 5.3). Intermetallic compound formation, here Cu_6Sn_5, at the bottom of the film, preferably along the (perpendicular) grain/crystal boundaries in the film, induces a negative stress-depth gradient in the film: parallel to the surface, the stress near the surface is positive (tensile) and negative (compressive) near the bottom; see the large, black arrows in the figure. This negative stress-depth gradient (induces a vacancy-concentration gradient and) tends to realize mass transport of tin from the bottom to the surface of the film, along the crystal/grain boundaries oriented more or less perpendicular to the film surface. At a location where a surface grain with inclined grain boundaries is present, such mass transport can be realized on a permanent basis, because shear stresses acting along these inclined grain boundaries (see the small black/white arrows in the figure; as a consequence of the planar state of stress in the film) can realize shear along these inclined grain boundaries, leading to outward growth of such surface grains, which thus become the whisker "roots" (adapted from *M. Sobiech, J. Teufel, U. Welzel, E.J. Mittemeijer and W. Hügel, Journal of Electronic Materials, 40 (2011), 2300–2313*)

boundaries within the Sn film (see Fig. 13.12). The positive volume misfit in the film, due to the development of IMC within the film, induces a compressive stress parallel to the tin/copper interface (i.e. parallel to the surface of the film) in the depth range of the IMC. At the same time a tensile stress (parallel to the surface of the film) occurs near the surface (where no IMC is present), which can be seen as the outcome of the realization/maintenance of mechanical equilibrium. The presence of a compressive (i.e. negative) stress near the bottom of the film and a tensile (i.e. positive) stress near the surface of the film implies that a *negative* stress-depth gradient exists in the film (*negative* because the stress decreases with increasing depth). It is this negative stress gradient that drives the whisker formation.[20]

---

[20] The negative stress-depth gradient evokes a vacancy-concentration depth gradient which causes a mass flux (of tin atoms) from the bottom part of the film to the top part of the film, likely via the grain boundaries at the low (room) temperature considered (a process that resembles the one known in materials science as "Coble creep"). Then, to maintain the massive nature of the film, whisker growth occurs, to accommodate the Sn transported to the film surface along the more or less perpendicular grain boundary.

The *negative nature* of the *stress gradient* as the "motor" for the whiskering is a fully new insight. Indeed, and in contrast with what was common sense among those investigating the whiskering phenomenon, we demonstrated in additional experiments that (i) even in the presence of a tensile stress in the surface region of the film, whiskering can occur provided the stress gradient is negative, and (ii) even in the presence of pronounced, compressive but homogeneous stress in the film (i.e. in the absence of a stress gradient), no whiskering occurs. Hence the original idea, that a compressive stress is necessary for whisker formation, is wrong.

This was the second time that I experienced, on the basis of our own research very many years later, that a generally believed understanding, as presented by Burgers in his lecture course, is, after all, untenable (for another such example, see earlier in this chapter under "*Interface and Surface Energetics; a Materials Science Approach; wrong "Diagnostics""*).

On the basis of the insight obtained, now various methods to inhibit whisker formation can be proposed. These derive from avoiding the development of a negative stress gradient, e.g. by obstructing IMC formation, or by annealing of the film causing stress relaxation, or by modifying the microstructure of the film such that very many crystal/grain boundaries inclined with respect to the surface occur *within* the film (a microstructure constituted of so-called "equiaxed" grains, i.e. a microstructure of grains of variable, irregular shape but on average of the same size in all directions): then gliding along grain boundaries (see footnote 20) becomes a global process in the film (in contrast with the films with columnar grains and only few grain boundaries inclined with respect to the surface), leading to overall stress relaxation, and whisker formation does not occur.

The materials scientist experiences great satisfaction if his/her research of processes occurring in and with materials leads to fundamental, physical understanding of nature that allows utilization in technologically important applications. This is an essential character trait of materials science.

Upon closer consideration it must be recognized that not every grain boundary in the film can act as a "root" for whisker formation. Only those (few) surface adjacent grains which have grain boundaries inclined which respect to the surface (and are directly above a perpendicular grain boundary aiding Sn transport from the bottom of the film; see Fig. 13.12) can give rise to whisker formation: only along the inclined grain boundaries of such a surface grain shear stresses act (as consequence of the planar state of stress in the film), which support the outward transport of tin along these inclined grain boundaries by gliding (i.e. shearing) along these grain boundaries. The surface grain with its inclined grain boundaries thus becomes the "whisker root".

14

Epilogue

Abstract The Farewell Lecture; the end of a scientific career.

The lecture hall in the Institute was packed. People were standing, leaning on the walls, and sitting on the ground next to the fully occupied rows of seats. This was formally the last lecture of the Phase Transformations Course. Eric looked into the audience. He felt pleased with what he saw. Apart from his last, current students, members of the Institute and the University, and former and present colleagues, he saw many of his past Ph.D. students and post-docs, who partly had even come from abroad to attend this lecture.

He began:

"Lange Zeit habe ich die Absicht gehabt „lautlos" in den Ruhestand zu entschwinden. Jedoch, von verschiedenen Seiten ist auf mich „eingewirkt" worden, doch eine „Abschiedsveranstaltung" zu erlauben und organisieren zu lassen. Als dann auch noch meine Frau, die auch meine Sekretärin ist, sich in diesem Chor eingeschaltet hat, war ich überzeugt, dass es doch sinnvoll wäre, nicht unbemerkt von der Bühne abzutreten. Was ich aber sicher nicht wollte, ist ein Abschiedssymposium: die dauern zu lang und für Nicht-Wissenschaftler sind sie unerträglich. So kam es, dass ich mich entschieden habe für eine „Abschiedsvorlesung" und das ist es, was Sie jetzt erleben werden.

Meine Abschiedsvorlesung trägt den Titel „Besonderheiten". Damit sind zum Teil „Besonderheiten" gemeint, die ich in meinem Leben als Wissenschaftler erfahren habe und wovon ich meine, dass sie Bedeutung nicht nur für mich

E. J. Mittemeijer, *How Science Runs,*
https://doi.org/10.1007/978-3-030-90095-3_14

haben. Das wird der erste, mehr persönlich gehaltene Teil des Vortrags sein, in dem ich Deutsch sprechen werde, unterstützt von Englischsprachigen Folien mit Schlüsselworten.[1]

In the second part of this lecture, where I will largely use the English language, I will focus on „particularities" of nature, as discovered and/or analyzed, mainly in recent years, by me and my department. These „particularities" serve to provide a picture of modern materials science, as a means to discover new and fundamental properties of nature."

Eric was familiar with the phenomenon "Farewell Lecture". It is a tradition at universities in The Netherlands and Anglo-Saxon countries. A few of these special lectures Eric had attended during his years in Delft. Usually these singular lectures were most enjoyable and also interesting for a general public. The retiring Professor is expected to give a lecture as if it were to be his last one. *That* really applied to Eric's Farewell Lecture. He said:

*"Diese Vorlesungsstunde in **diesem Hörsaal** an **diesem Tag** zu **diesem Zeitpunkt** ist tatsächlich die planmässig letzte Vorlesungsstunde der Vorlesung Phase Transformations/Phasenumwandlungen für Masterstudenten Materialwissenschaft, die ich zum letzten Mal in diesem Semester damit abschliesse. Nur ist der Inhalt dieser Vorlesung in dieser letzten Stunde natürlich ein ganz Anderer als dies in den vergangenen Jahren der Fall war.*

*"This lecture hour in **this auditorium** on **this day** at **this time** is actually the scheduled last lecture hour of the lecture course Phase Transformations for Master students Materials Science, which I will thereby conclude a last time in this semester. Only the content of this lecture in this last hour is of course very different from what it was in previous years."*

In Germany this type of farewell ceremony is not, or at least much less, well known. Here often a one-day symposium is organized, where most presentations are just about scientific developments related to the science of the retiring Professor. As indicated with the introduction to his farewell lecture, Eric found such events less fitting to the occasion.

---

[1] The translation into English of this German text runs about as follows:

"For a long time I had the intention to disappear "noiselessly" into retirement. However, from several sides people exerted "pressure" onto me yet to agree with and let organize a "farewell event". If then my wife, who is also my secretary, became part of this choir too, I became convinced of the meaningfulness not to vanish unnoticed from the stage of science. However, a farewell symposium was certainly not what I wanted: they take too long and are unbearable for non-scientists. So it was that I decided to give a "Farewell Lecture" and that it is what you are about to experience.

My Farewell Lecture bears the title "Particularities". With this notion partly those "particularities" are meant which I have experienced during my life as scientist and of which I believe that they have a bearing not only for me. This will concern the first, more personal part of my lecture, where I will speak German with support of transparencies with keywords in English."

In contrast with very many years ago, Eric had no stage fright before he delivered a lecture for whichever gathering, conference or symposium, at whatever location on the world. But this special time he experienced a feeling of tension: he was uncertain if this lecture would be interesting at best appealing and hopefully to some extent charming for the present audience of such hugely diverse background. After he had spoken his opening statements, reproduced above, calmness came down on him. He felt the audience followed him in what he said.

The preceding year had taken a completely different course from what possibly could have been expected at its start. Because Eric had extended his professional career beyond the age of 65, his wife and secretary Marion, of the same age as Eric, had done the same, in order to guarantee Eric continued secretarial support as he was used to; they were a very good team. During this last year they had planned to move from Stuttgart to their new apartment in Heidelberg. About a week before this move, Marion and Eric were already in a progressed stage of packing, it became suddenly clear, in a dramatic way, that Marion was severely ill. On the day of the move to Heidelberg, they "in-between" also went to a specialist at the University Clinic of Heidelberg and it was disclosed to them that an immediate, large operation was unavoidable. Not many weeks later the Farewell Lecture was scheduled to take place. It had been widely announced and cancelling it was no real option. The treating physician said it was possible that Marion might still attend Eric's Farewell Lecture. And so it came, that Eric was sitting for days, at a small table at the foot of the bed of Marion in the hospital, behind his laptop, preparing his Farewell Lecture, while a large part of the boxes of the move to Heidelberg remained unpacked. And now Marion was there, sitting with difficulty in the front row, shortly after her release from hospital, still with pain only partly suppressed by pain killers. Eric avoided looking at her during the lecture; it would stir emotions he feared not being able to control.

The first part of the lecture, devoted to a number of also charming anecdotes revealing human "particularities", taken from Eric's personal experiences as a scientist, were framed by two sayings creating an area of tension between them: "*Those who only speak about the past have no future*" and, to "shield" Eric for such criticism: "*A successful future requires a clear remembrance*" (this last assertion, in this form ("*Gute Zukunft braucht klare Erinnerung*"), is due to Richard von Weizsäcker). Then, in the second part of the lecture, "baffling" phenomena, which recently had occupied the mind of the materials scientist, were recounted such that they in principle could fascinate the layman as well and at the same time would give an impression of recent advances in materials science.

Having arrived near to the end of his presentation, Eric felt tired. He hinted at a phrase from Pablo Casals, a world famous cello player: "*A man who loves his job never ages*". He paused a moment and then said: "*This slogan is untrue*". The audience laughed. He said he felt the loss of energy that had occurred slowly but noticeably during the years and that men and women would be wise, upon becoming older, to accept that the quantitative level of power needed to continue in the same way as before could only be brought up with ever increasing effort. One should know when to go. He loved his job greatly, but felt it time to retire, not to become mute in the world of science, but to adapt his activity to changed boundary conditions.

Then Eric explicitly thanked the Max Planck Society for having offered to him these extraordinary, unique possibilities for developing his research, leading to the best, most successful research years of his life here in Stuttgart at the Institute. And he thanked the University of Stuttgart, in particular the Faculty of Chemistry, under which care the materials science, especially as an academic course, had flourished.

Finally, the most difficult part of his Farewell Lecture had to be mastered. Eric wanted to thank Marion and knew he had to fight with his voice, as this, especially at this time, had a deep meaning to both of them. Somehow he managed to keep control, although his voice moved to a revealing, lower tone and different timbre. Marion was unprepared and yet responded wonderfully.

After a short break Professor Joachim Spatz of the Institute and Professors Thomas Schleid and Guido Schmitz from the Faculty spoke illuminating Eric's career. The reception thereafter was a great delight for Eric and Marion, meeting so many former students and co-workers, not least the colleagues and friends from especially China, Denmark and Italy.

Later, everybody gone, Eric and Marion together sat in Eric's office, a last time; tired, satisfied and somewhat melancholic. Then they drove to Heidelberg. A scientific career had come to its end.

Printed in the United States
by Baker & Taylor Publisher Services